D0372346

# UNDERSTANDING
# ARTIFICIAL
# INTELLIGENCE

# UNDERSTANDING ARTIFICIAL INTELLIGENCE

FROM THE EDITORS OF *SCIENTIFIC AMERICAN*

Compiled and with introductions by

Sandy Fritz

Foreword by Rodney Brooks

A Byron Preiss Book

**WARNER BOOKS**

An AOL Time Warner Company

The following essays first appeared in the pages of *Scientific American* magazine, as follows: "On Computational Wings: Rethinking the Goals of Artificial Intelligence" by Kenneth M. Ford and Patrick J. Hayes, 1998; "2001: A Scorecard" by Gary Stix, January 2001; "Artificial Intelligence" by Douglas B. Lenat, September 1995; "Fuzzy Logic" by Bart Kosko, July 1993; "How Neural Networks Learn from Experience" by Geoffrey E. Hinton, September 1992; "Computing with Molecules" by Mark A. Reed and James M. Tour, June 2000; "Intelligent Materials" by Craig A. Rogers, September 1995; "The Coming Merging of Mind and Machine" by Ray Kurzweil, Special Issue: Your Bionic Future; "Rise of the Robots" by Hans Moravec, December, 1999; "Will Robots Inherit the Earth?" by Marvin Minksy, October 1994; "The Mastermind of Artificial Intelligence" by John Horgan, November 1993.

Warner Books, Inc., 1271 Avenue of the Americas, New York, NY 10020

Visit our Web site at www.twbookmark.com.

An AOL Time Warner Company

Printed in the United States of America

ISBN: 0-446-67875-9
Library of Congress Control Number: 2001094090

Cover design by J. Vita
Book design by Gilda Hannah

# Contents

Rodney Brooks

S tarting with Alan Turing, artificial intelligence (AI) research has been a driving force behind much of computer science for over fifty years. Turing wanted to build a machine that was as intelligent as a human being since it was possible to build imitations of "any small part of a man." He suggested that instead of producing accurate electrical models of nerves, a general-purpose computer could be used and the nervous system could be modeled as a computational system. He suggested that "television cameras, microphones, loudspeakers," etc., could be used to model the rest of a humanoid body. "This would be a tremendous undertaking of course," he acknowledged. Even so, Turing noted that the so constructed machine " . . . would still have no contact with food, sex, sport and many other things of interest to the human being." Turing concluded that in the 1940s the technical challenges to building such a robot were too great and that the best domains in which to explore the mechanization of thought were various games, and cryptoanalysis, "in that they require little contact with the outside world." He explicitly worried about ever teaching a computer a natural human language as it "seems however to depend rather too much on sense organs and loco-motion to be feasible."

Turing thus set out the format for early artificial intelligence research, and in doing so touched on many of the issues that are still the points of hot debate in 2001, and which are discussed in many of the chapters of this book. Much of the motivation for artificial intelligence is the inspiration from people—that they can walk, talk, see, think, and do.

How can we make machines that can do these things too?

The first issue to be resolved is whether people are some-how intrinsically different from machines. One side argues that just as we had to adapt to not living at the center of the universe, then had to adapt to having evolved from animals, now we will have to adapt to being no more special than com-plicated machines. Others argue that there is something spe-cial about being human and mere machines can never have the capabilities or the personhood of humans.

The second issue is whether our intelligence is something that can be emulated computationally. Some argue that the brain is an information-processing machine, made of meat, and as such can be replaced by a fast computer—and Moore's law is ensuring that we will have a fast enough computer within the next 20 years. Others argue that perhaps there is something non-computational going on inside our heads—not necessarily anything that is beyond the understanding of cur-rent physics, but that the organization of whatever is going on is not yet understood even at a basic level.

The third issue is how the computation, or whatever it is, should be organized. Are we the product of rational thought, or are we rather dressed up animals, and the product of reactive brains programmed by evolution to fight or flee?

And the fourth issue is how to get all the necessary capabil-ities into a machine. Can they be explicitly written down as rules and be digested by a disembodied computer? Or, do we need to build robots with sensors and actuators that live in the world and learn what they need to know from their interactions with the world?

Finally there are speculations on where our work on AI will lead. Such speculations have been part of the field since its earliest days, and will continue to be part of the field. The details of the speculations usually turn out to be wrong, but the questions they raise are often profound and important, and are ones we all should think about.

have actually surpassed
r C. Clarke's and film d
ey Kubrick's vision of c
ology at the turn of the
's computers are vastly
er, more portable and us
faces the typ
l controls found on the
very 1. But by and large

# Introduction

## Sandy Fritz

**O**f all the machines that have changed our lives, perhaps the most influential is the computer, the quintessential child of the industrial revolution. It is mass produced and easily available. It combines brilliant scientific insight with a product that improves the real world. The speed of its growth is astounding. Its refinement and improvement rides the very edge of scientific achievement. Computers can generate realistic images of dinosaurs on the run, defeat human chess champions, and serve as a humble hubs linking anybody in the world with the World Wide Web. Truly impressive.

But can it think? If a machine can think for itself on some level, perhaps even learn from and improve upon its performance through experience, it can be a far more useful tool. Just about everyone agrees this is a desirable thing, but how to get there is a whole other question.

One camp in the artificial intelligence field sees the human brain as a computer that can be copied to produce an artificial mind. Others argue that human behavior defies the strictures of a computer program. At the heart of this discussion lies the truly slippery question facing those who would fashion artificial intelligence: What does it mean to think? And its corollary: Does thinking constitute consciousness?

In the attempt to get computers to behave in a somewhat thinking/conscious manner, scientists have had to study human beings in a whole new light. It turns out that much of what we take for granted in the world—that a dropped apple will fall to the ground or that rain makes things wet—constitutes an invisible set of assumptions that form a backdrop to all human interactions with the world. Thus, before an intelligent robot maid can vacuum the floor, a mass of facts about the 3-D world—the amount of pressure needed to remove dirt from a carpet; the difference between an ink stain and a black-and-white design, etc.—must be coded and downloaded.

Human beings learn by integrating their sensory experiences of the world into patterns. We do it naturally, automatically, incorporating and cross-referencing thoughts, conversation and interactions with the world at large. A single interaction with an apple or a snowstorm or a car that won't start brings us reams of new information about the world—processed without conscious effort and seamlessly integrated with what we already know to create a better, broader understanding.

Neural networks, designed to promote and create AI, approach the *experience-learning-new information-integration* problem by modeling the biology of the human nervous system. These vastly simplified networks show promise, but the computing cycles necessary to power them make complicated neural nets run slowly.

The speed breakthrough that foreshadows the jump to true AI may hinge in the switch from silicon memory to molecular memory (as made clear by Mark A. Reed and James M. Tour in "Computing with Molecules"). Combine that with the ability of molecules to function as switches and other products of nanotechnology, and the accompanying increase in processing power could set the stage for AI's grand show: the intelligent robot.

The dream of a mobile, autonomous, sentient, non-biological creation has been with us for about a century. Some predict the robot will be a conscious entity, capable of feelings, emotions, and insights. Some cast the robot as a task-specific machine, thoughtful perhaps, but empty of human considerations and subtleties.

Robots, and certain other limited forms of artificial intelligence, have made their mark in the world already, from supercomputers to robotic factories. The questions and challenges that face the latest generation of researchers in this field, explored here in detail, make it clear that we are well on the way to thinking machines becoming an important player in the details of everyday life.

*The Turing test for consciousness shaped the early efforts in the artificial intelligence field. Can a machine convince a human being that it, the machine, is human? Perhaps the test is flawed and should be discarded, say authors Ford and Hayes. The greatest value of artificial intelligence may lie not in imitating human thinking but in extending it into new realms.*

# On Computational Wings: Rethinking the Goals of Artificial Intelligence

## Kenneth M. Ford and Patrick J. Hayes

**M**any philosophers and humanist thinkers are convinced that the quest for artificial intelligence has turned out to be a failure. Eminent critics have argued that a truly intelligent machine cannot be constructed and have even offered mathematical proofs of its impossibility. And yet the field of artificial intelligence is flourishing. "Smart" machinery is part of the information-processing fabric of society, and thinking of the brain as a "biological computer" has become the standard view in much of psychology and neuroscience.

While contemplating this mismatch between the critical opinions of some observers and the significant accomplishments in the field, we have noticed a parallel with an earlier endeavor that also sought an ambitious goal and for centuries was attacked as a symbol of humankind's excessive hubris: artificial flight. The analogy between artificial intelligence and artificial flight is illuminating. For one thing, it suggests that the traditional view of the goal of AI—to create a machine that can successfully imitate human behavior—is wrong.

For millennia, flying was one of humanity's fondest dreams.

The prehistory of aeronautics, both popular and scholarly, dwelled on the idea of imitating bird flight, usually by somehow attaching flapping wings to a human body or to a framework worn by a single person. It was frustratingly clear that birds found flying easy, so it must have seemed natural to try to capture their secret. Some observers suggested that bird feathers simply possessed an inherent "lightness." Advocates of the possibility of flight argued that humans and birds were fundamentally similar, whereas opponents argued that such comparisons were demeaning, immoral or wrongheaded. But both groups generally assumed that flying meant imitating a bird. Even relatively sophisticated designs for flying machines often included some birdlike features, such as the beak on English artist Thomas Walker's 1810 design for a wooden glider.

This view of flying as bird imitation was persistent. An article in English Mechanic in 1900 insisted that "the true flying machine will be to all intents and purposes an artificial bird." A patent application for a "flying suit" covered with feathers was made late in the 19th century, and wing-flapping methods were discussed in technical surveys of aviation published early in this century.

## The Turing Test

Intelligence is more abstract than flight, but the long-term ambition of AI has also traditionally been characterized as the imitation of a biological exemplar. When British mathematician Alan M. Turing first wrote of the possibility of artificial intelligence in 1950, he suggested that AI research might focus on what was probably the best test for human intelligence available at the time: a competitive interview. Turing suggested that a suitable test for success in AI would be an "imitation game" in which a human judge would hold a three-way conversation with a computer and another human and try to tell them apart. The judge would be free to turn the conversation to any topic, and the successful machine would be able to chat about it as convincingly as the human. This would require the machine participant in the game to understand language and conversational conventions and to have a general ability to reason. If the judge could not tell the difference after some reasonable amount of time, the machine would pass the test: it would be able to seem human to a human.

There is some debate about the exact rules of Turing's imitation game, and he may not have intended it to be taken so seriously. But some kind of "Turing test" has become widely perceived, both inside and outside the field, as the ultimate goal of artificial intelligence, and the test is still cited in most textbooks. Just as with early thinking about flight, success is defined as the imitation of a natural model: for flight, a bird; for intelligence, a human.

The Turing test has received much analysis and criticism, but we believe that it is worse than often realized. The test has led to a widespread misimpression of the proper ambitions of our field. It is a poorly designed experiment (depending too much on the subjectivity of the judge), has a questionable technological objective (we already have lots of human intelligence) and is hopelessly culture-bound (a conversation that is passable to a British judge might fail according to a Japanese or Mexican judge). As Turing himself noted, one could fail the test by being too intelligent—for example, by doing mental arithmetic extremely fast. According to media reports, some judges at the first Loebner competition in 1991-a kind of Turing test contest held at the Computer Museum in Boston-rated a human as a machine on the grounds that she produced extended, well-written paragraphs of informative text. (Apparently, this is now considered an inhuman ability in parts of our culture.) With the benefit of hindsight, it is now evident that the central defect of the test is its species-centeredness: it assumes that human thought is the final, highest pinnacle of thinking against which all others must be judged. The Turing test does not admit of weaker, different or even stronger forms of intelligence than those deemed human.

Most contemporary AI researchers explicitly reject the goal of the Turing test. Instead they are concerned with exploring the computational machinery of intelligence itself, whether in humans, dogs, computers or aliens. The scientific aim of AI research is to understand intelligence as computation, and its engineering aim is to build machines that surpass or extend human mental abilities in some useful way. Trying to imitate a human conversation (however "intellectual" it may be) contributes little to either ambition.

In fact, hardly any AI research is devoted to trying to pass the Turing test. It is more concerned with issues such as how machine learning and vision might be improved or how to design an autonomous spacecraft that can plan its own actions. Progress in AI is not measured by checking fidelity to a human conversationalist. And yet many critics complain of a lack of progress toward this old ambition. We think the Turing test should be relegated to the history of science, in the same way that the aim of imitating a bird was eventually abandoned by the pioneers of flight. Beginning a textbook on AI with the Turing test (as many still do) seems akin to starting a primer on aeronautical engineering with an explanation that the goal of the field is to make machines that fly so exactly like pigeons that they can even fool other pigeons.

## Imitation vs. Understanding

Researchers in the field of artificial intelligence may take a useful cue from the history of artificial flight. The development of aircraft succeeded only when people stopped trying to imitate birds and instead approached the problem in new ways, thinking about airflow and pressure, for example. Watching hovering

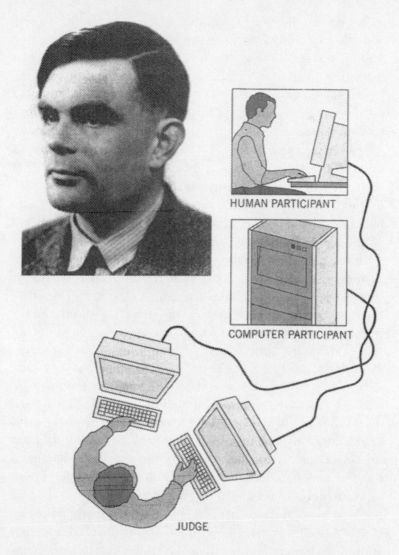

HUMAN PARTICIPANT

COMPUTER PARTICIPANT

JUDGE

The Turing test for artificial intelligence was proposed in 1950 by British mathematician Alan M. Turing (photo top left). In the test, a judge would hold a three-way conversation with a computer and another human. If the judge cannot distinguish between the responses of the human and those of the computer, the machine would pass the test.

gulls inspired the Wright brothers to use wing warping—turning an aircraft by twisting its wings—but they did not set out to imitate the gull's wing. Starting with a box kite, they first worked on achieving sufficient lift, then on longitudinal and lateral stability, then on steering and finally on propulsion and engine design, carefully solving each problem in turn. After that, no airplane could be confused with a bird either in its overall shape or in its flying abilities. In some ways, aircraft may never match the elegant precision of birds, but in other ways, they outperform them dramatically. Aircraft do not land in trees, scoop fish from the ocean or use the natural breeze to hover motionless above the countryside. But no bird can fly at 45,000 feet or faster than sound.

Rather than limiting the scope of AI to the study of how to mimic human behavior, we can more usefully construe it as the study of how computational systems must be organized in order to behave intelligently. AI programs are often components of larger systems that are not themselves labeled "intelligent." There are hundreds of such applications in use today, including those that make investment recommendations, perform medical diagnoses, plan troop and supply movements in warfare, schedule the refurbishment of the space shuttle and detect fraudulent use of credit cards. These systems make expert decisions, find meaningful patterns in complex data and improve their performances by learning. All these actions, if done by a human, would be taken to display sound judgment, expertise or responsibility. Many of these tasks, however, could not be done by humans, who are too slow, too easily distracted or not sufficiently reliable. Our intelligent machines already surpass us in many ways. The most useful computer applications, including AI applications, are valuable exactly by virtue of their lack of humanity. A truly humanlike program would be just as useless as a truly pigeonlike aircraft.

## Waiting for the Science

The analogy with flight provides another insight: technological advances often precede advances in scientific knowledge. The designers of early aircraft could not learn the principles of aerodynamics by studying the anatomy of birds. Evolution is a sloppy engineer, and living systems tend to be rich with ad hoc pieces of machinery with multiple uses or mechanisms jury-rigged from structures that evolved earlier for a different reason. As a result, it is often very difficult to discover basic principles by imitating natural mechanisms.

Experimental aerodynamics became possible only in the early part of this century, when artificial wings could be tested systematically in wind tunnels. It did not come from studying natural exemplars of flight. That a gull's wing is an airfoil is now strikingly obvious, yet the airfoil was not discovered by examining the anatomy of birds. Even the Wright brothers never really understood why their Flyer flew. The aerodynamic principles of the airfoil emerged from experiments done in 1909 by French engineer Alexandre-Gustave Eiffel, who used a wind tunnel and densely instrumented artificial wings. The first aircraft with "modern" airfoils—which were made thicker after engineers demonstrated that thicker airfoils improved lift without increasing drag—did not appear until late in World War I. As is true for many other disciplines, a firm theoretical understanding was possible only when controlled experiments could be done on isolated aspects of the system. Aerodynamics was discovered in the laboratory.

The same reasoning applies to the study of human intelligence. It may be impossible to discover the computational principles of intelligent thought by examining the intricacies of human thinking, just as it was impossible to discover the principles of aerodynamics by examining bird wings. The Wright brothers' success was largely attributed to their perception of flight in terms of lift, control and power; similarly, a science of intelligence must isolate particular aspects of thought, such as memory, search and adaptation, and allow us to experiment on these one at a time using artificial systems. By systematically varying functional parameters of thought, we can determine the ways in which various kinds of mental processes can interact and support one another to produce intelligent behavior.

Several areas of AI research have been transformed in the past decade by an acceptance of the fact that progress must be measurable, so that different techniques can be objectively compared. For example, large-scale empirical investigations must be conducted to evaluate the efficiency of different search techniques or reasoning methods. In this kind of AI research, computers are providing the first wind tunnels for thought.

## A Science of Intelligence

Rejecting the Turing test may seem like a retreat from the grand old ambition of creating a "humanlike" mechanical intelligence. But we believe that the proper aim of AI is much larger than simply mimicking human behavior. It is to create a computational science of intelligence itself, whether human, animal or machine. This is not a new claim; it has been made before by AI pioneers Allen Newell and Herbert A. Simon, cognitive psychologist Zenon Pylyshyn and philosopher Daniel C. Dennett, among others. But it was not until we noted the analogy with artificial flight that we appreciated the extent to which the Turing test, with its focus on imitating human performance, is so directly at odds with the proper objectives of AI. Some of our colleagues say their ultimate goal is indeed the imitation of human intelligence. Even with this limited aim, however, we believe that the perspective sketched here provides a more promising way to achieve that ambition than does the method outlined by Turing.

Consider again the analogy with flight. Just as the principles of aerodynamics apply equally to any wing, natural or artificial, the computational view of intelligence—or, more broadly, of mentality—applies just as well to natural thinkers as to artificial thinkers. If cognitive psychology and psycholinguistics are like the study of bird flight in all its complexity, then applied AI is like aeronautical engineering. Computer science supplies the principles that guide the engineering, and computation itself is the air that supports the wings of thought.

The study of artificial intelligence, like a large part of computer science, is essentially empirical. To run a program is often to perform an experiment on a large, complex apparatus (made partly of metal and silicon and partly of symbols) to discover the laws that relate its behavior to its structure. Like artificial wings, these AI systems can be designed and instrumented to isolate particular aspects of this relation. Unlike the research methodology of psychology, which employs careful statistical analysis to discern relevant aspects of behavior in the tangled complexity of nature, the workings of AI systems are open to direct inspection. Using computers, we can discover and experiment directly with what Newell and Simon have called the "laws of qualitative structure."

This picture of AI defines the field in a more useful and mature way than Turing could provide. In this view, AI is the engineering of cognitive artifacts based on the computational understanding that runs through and informs current cognitive science. Turing correctly insisted that his test was not meant to define intelligence. Nevertheless, in giving us this touchstone of success, he chose human intelligence—in fact, the arguing skill of an educated, English middle-class man playing a kind of party game—as our goal. But the very science that Turing directed us toward provides a perspective from which a much broader and more satisfying account of intelligence is emerging.

## Scholastic Critics

Artificial intelligence and artificial flight are similar even in the criticisms they attract. The eminent American astronomer

Simon Newcomb argued passionately in the early 1900s against the idea of heavier-than-air flight. Newcomb's fulminations seem amusing now, but his arguments were quite impressive and reflected the view of the informed intelligentsia of his day. Like British mathematical physicist Roger Penrose, who uses Gödel's theorem to "prove" that AI is impossible, Newcomb employed mathematical arguments. He pointed out that as birds get bigger, their wing area increases in proportion to the square of their size, but their body weight increases in proportion to the cube, so a bird the size of a man could not fly. He was still using this argument against the possibility of manned flight several years after the Wright brothers' success at Kitty Hawk, N.C., when aircraft were regularly making trips lasting several hours. It is, in fact, quite a good argument—aircraft takeoff weights are indeed roughly proportional to the cube of their wingspan—but Newcomb had no idea how sharply the lift from an airfoil increases in proportion to its airspeed. He thought of a wing as simply a flat, planar surface.

Newcomb also used a combination of thought experiment and rhetoric to make his point—the same tactic that philosopher John R. Searle has employed in his famous "Chinese Room" argument against AI. Newcomb stated scornfully, "Imagine the proud possessor of the aeroplane darting through the air at a speed of several hundred feet per second! It is the speed alone that sustains him. How is he ever going to stop?" Newcomb's arguments, with their wonderful combination of energy, passion, cogency and utter wrongheadedness, are so similar to contemporary arguments against artificial intelligence that for several years we have offered the annual Simon Newcomb Award for the silliest new argument attacking AI. We welcome nominations.

A common response to our analogy between artificial intelligence and artificial flight is to ask what will be the Kitty Hawk of AI and when will it happen. Our reply follows that of Herbert Simon: it has already happened. Computers regularly perform intelligent tasks and have done so for many years. Artificial intelligence is flying all around us, but many simply refuse to see it. Among the thousands of applications in use today, here are just a few examples: AI systems now play chess, checkers, bridge and backgammon at world-class levels, compose music, prove mathematical theorems, explore active volcanoes, synthesize stock-option and derivative prices on Wall Street, make decisions about credit applications, diagnose motor pumps, monitor emulsions in a steel mill, translate technical service manuals, and act as remedial reading tutors for elementary school children. In the near future, AI applications will guide deep-space missions, explore other planets and drive trucks along freeways.

But should all this really count as "intelligent"? The performance of AI systems, like the speed or altitude of aircraft, is not open to dispute, but whether or not one chooses to call it "intelligent" is determined more by social attitude than by anything objective. When any particular ability is mechanized, it is often no longer considered to be a hallmark of mental prowess. It is easy now to forget that when Turing was writing, a "computer" was a human being who did arithmetic for a living, and it was obvious to everyone that computing required intelligence. The meaning of the word has now changed to mean a machine, and performing fast, accurate arithmetic is no longer considered a hallmark of mental ability, just as the meaning of "flying" has changed to cover the case, once inconceivable, of dozing quietly in an airplane seat while traveling at hundreds of miles an hour far above the clouds. Newcomb—who was famous as one of the finest computers of his time—went to his deathbed refusing to concede that what early aircraft did should be called "flying."

Turing suggested his test as a way to avoid useless disputes about whether a particular task counted as truly intelligent. With considerable prescience, he anticipated that many people would never accept that the action of a machine could ever be labeled as "intelligent," that most human of labels. But just as there was no doubt that the early flyers moved through the air at certain altitudes and speeds, there is no doubt that electronic computers actually get arithmetic done, make plans, produce explanations and play chess. The labels are less important than the reality.

The arbitrariness of the social labeling can be illustrated by a thought experiment in which the machine is replaced by something mysterious but natural. Whereas a dog will never pass the Turing test, no one but a philosopher would argue that a dog does not display some degree of intelligence—certainly no one who has owned a dog would make such an argument. It is often claimed that Deep Blue, the computer that defeated chess champion Garry Kasparov, is not really intelligent, but imagine a dog that played chess. A chess-playing dog that could beat Kasparov would surely be acclaimed a remarkably smart dog.

The idea that natural intelligence is a complex form of computation can only be a hypothesis at present. We see no clear reason, however, why any mental phenomenon cannot be accounted for in this way. Some have argued that the computationalist view cannot account for the phenomenology of consciousness. If one surveys the current theories of the nature of consciousness, however, it seems to us that a computationalist account offers the most promise. Alternative views consider consciousness to be some mysterious physical property, perhaps arising from quantum effects influenced by the brain's gravity or even something so enigmatic as to be forever beyond the reach of science. None of these views seems likely to explain how a physical entity, such as a brain in a body, can come to be aware of the world and itself. But the AI view of mental life as the product of computation provides a detailed account of how internal symbols can have meaning for the machine and how this meaning can influence and be influenced by the causal relations between the machine and its surroundings.

The scientific goal of AI is to provide a computational account of intelligence or, more broadly, of mental ability itself—not merely an explanation of human mentality. This very understanding, if successful, must deny the uniqueness of human thought and thereby enable us to extend and amplify it. Turing's ultimate aim, which we can happily share, was not to describe the difference between thinking people and unthinking machines but to remove it. This is not to disparage or reduce humanity and still less to threaten it. If anything, understanding the intricacies of airflow increases our respect for how extraordinarily well birds fly. Perhaps it seems less magical, but its complexity and subtlety are awesome. We suspect that the same will be true for human intelligence. If our brains are indeed biological computers, what remarkable computers they are.

*Fads come and fads go, but pop-cultural icons have staying power. Certain popular images can inform the core assumptions of a generation. HAL 9000, the sentient computer of 2001: A Space Odyssey, is one such. Arthur C. Clarke and Stanley Kubrick meant for HAL—and the entire starship Discovery, shaped like an artificial central nervous system with HAL as the brain—to be a metaphor for humanity's next steps in exploration. But an entire generation, or two, has taken it to be the blueprint for evolution of intelligent machines. With the passage of three-plus decades, it is both amusing and informative to look back on the film and see just how wonderful the future was meant to be, and how far we've come to realizing it.*

# 2001: A Scorecard

## Gary Stix

I t will always be easier to make organic brains by unskilled labor than to create a machine-based artificial intelligence. This joke about doing things the old-fashioned way, which appears in the book version of *2001: A Space Odyssey,* still has an undeniable ring of truth. The science-fiction masterpiece will be remembered best for the finely honed portrait of a machine that could not only reason but also experience the epitome of what it means to be human: neurotic anxiety and self-doubt.

The Heuristically programmed ALgorithmic Computer, a.k.a. HAL, may serve as a more fully rounded representation of a true thinking machine than the much vaunted Turing test, in which a machine proves its innate intelligence by fooling a human into thinking that it is speaking to one of its own kind. (See "On Computational Wings," page 5.) In this sense, HAL's abilities— from playing chess to formulating natural speech and reading lips—may serve as a better benchmark for measuring machine smarts than a computer that can spout vague, canned maxims that a human may interpret as signs of native intelligence.

Surprisingly, perhaps, computers in some cases have actually surpassed writer Arthur C. Clarke's and film director Stanley Kubrick's vision of computing technology at the turn of the millennium. Today's computers are vastly smaller, more portable and use software interfaces that forgo the type of manual controls found on the spaceship *Discovery 1*. But by and large, computing technology has come nowhere close to HAL. David G. Stork, who edited *HAL's Legacy: 2001's Computer as Dream and Reality*, a collection of essays comparing the state of computing with HAL's capabilities, remarks that for some defining characteristics of intelligence—language, speech recognition and understanding, common sense, emotions, planning, strategy, and lip reading—we are incapable of rendering even a rough facsimile of a HAL. "In all of the human-type problems, we've fallen far, far short," Stork says.

Even computer chess, in which seeming progress has been made, deceives. In 1997 IBM's Deep Blue beat then world champion Garry Kasparov. Deep Blue's victory, though, was more a triumph of raw processing power than a feat that heralded the onset of the age of the intelligent machine. Quantity had become quality, Kasparov said in describing Deep Blue's ability to analyze 200 million chess positions a second. In fact, Murray F. Campbell, one of Deep Blue's creators, notes in *HAL's Legacy* that although Kasparov, in an experiment, sometimes failed to distinguish between a move by Deep Blue and one of a human grandmaster, Deep Blue's overall chess style did not exhibit human qualities and therefore was not "intelligent." HAL, in contrast, played like a real person. The computer with the unblinking red eye seemed to intuit from the outset that its opponent, *Discovery* crewman Frank Poole, was a patzer [inept player], and so it adjusted its strategy accordingly. HAL would counter with a move that was not the best one possible, to draw Poole into a trap, unlike Deep Blue, which assumes that its opponent always makes the strongest move and therefore counters with an optimized parry.

The novel of *2001* explains how the HAL 9000 series developed out of work by Marvin Minsky of the Massachusetts Institute of Technology and another researcher in the 1980s that showed how "neural networks could be generated automatically—self-replicated—in accordance with an arbitrary learning program [see "How Neural Networks Learn from Experience," page 43]. Artificial brains could be grown by a process strikingly analogous to the development of the human brain." Ironically, Minsky, one of the pioneers of neural networks who was also an adviser to the filmmakers (and who almost got killed by a falling wrench on the set), says today that this approach should be relegated to a minor role in modeling intelligence, while criticizing the amount of research devoted to it. "There's only been a tiny bit of work on commonsense reasoning, and I could almost characterize the rest as various sorts of get-rich-quick schemes, like genetic algorithms [and neural networks] where you're hoping you won't have to figure anything out," Minsky says.

In any case, Clarke remains undeterred by how far off the mark his vision has strayed. Machine intelligence will become more than science fiction, he believes, if not by this year. "I think it's inevitable; it's just part of the evolutionary process," he says. Errors in prediction, Clarke maintains, get counterbalanced over time by outcomes more fantastic than the original insight. "First our expectations of what occurs outrun what's actually happening, and then eventually what actually happens far exceeds our expectations."

Quoting himself (Clarke's Law), Clarke remarks that "any sufficiently advanced technology is indistinguishable from magic; as technology advances it creates magic, and [AI is] going to be one of them." Areas of research that target the ultimate in miniaturization, he adds, may be the key to making good minds. "When nanotechnology is fully developed, they're going to churn [artificial brains] out as fast as they like." Time will tell if that's prediction, like Clarke's speculations about telecommunications satellites, or just a prop for science fiction.

*Our experience of the world is built upon assumptions that seem simple and obvious to humans. We learn that a brisk half-mile walk feels good when the weather is temperate, not so good when the temperature is below freezing, and downright dangerous in the middle of a blizzard or hurricane. Such assumptions, learned from experiencing the real world, collectively fall into the category of common sense. It is these intuitive understandings of the world and people that prove to be most problematic in codifying for machines, but without them no machine could be thought of as truly "intelligent."*

# Programming Artificial Intelligence

## Douglas B. Lenat

O ne of the most frustrating lessons computers have taught us time and time again is that many of the actions we think of as difficult are easy to automate—and vice versa. In 1944 dozens of people spent months performing the calculations required for the Manhattan Project. Today the technology to do the same thing costs pennies. In contrast, when researchers met at Dartmouth College in the summer of 1956 to lay the groundwork for artificial intelligence, none of them imagined that 40 years later we would have come such a short distance toward that goal.

Indeed, what few successes AI has had point out the weakness of computerized reasoning as much as they do its narrow strengths. In 1965, for example, Stanford University's Dendral project automated sophisticated reasoning about chemical structures; it generated a list of all the possible three-dimensional structures for a compound and then applied a small set of simple rules to select the most plausible ones. Similarly, in 1975 a program called Mycin surpassed the average physician in the accuracy with which it diagnosed meningitis in patients.

It rigorously applied the criteria that expert clinicians had developed over the years to distinguish among the three different causes of the disease. Such tasks are much better suited to a computer than to a human brain because they can be codified as a relatively small set of rules to follow; computers can run through the same operations endlessly without tiring.

Meanwhile many of the tasks that are easy for people to do—figure out a slurred word in a conversation or recognize a friend's face—are all but impossible to automate, because we have no real idea of how we do such things. Who can write down the rules for recognizing a face?

As a result, amid the explosive progress in computer networking, user interface agents and hardware, artificial intelligence appears to be an underachiever. After initial gains led to high expectations during the late 1970s and early 1980s, there was a bitter backlash against AI in both industry and government. Ironically, in 1984, just as the mania was at its peak, I wrote an article for *Scientific American* in which I dared to be fairly pessimistic about the coming decade. And now that the world has all but given up on the AI dream, I believe that artificial intelligence stands on the brink of success.

My dire predictions arose because the programs that fueled AI hype were not savants but idiot savants. These so-called expert systems were often right, in the specific areas for which they had been built, but they were extremely brittle. Given even a simple problem just slightly beyond their expertise, they would usually get a wrong answer, without any recognition that they were outside their range of competence. Ask a medical program about a rusty old car, and it might blithely diagnose measles.

Furthermore, these programs could not share their knowledge. Mycin could not talk to programs that diagnosed lung diseases or advise doctors on cancer chemotherapy, and none of the medical programs could communicate with expert scheduling systems that might try to allocate hospital resources. Each represented its bit of the world in idiosyncratic and incompatible ways because developers had cut corners by incorporating many task-specific assumptions. This is still the case today.

## No Program Is an Island

People share knowledge so easily that we seldom even think about it. Unfortunately, that makes it all the more difficult to build programs that do the same. Many of the prerequisite skills and assumptions have become implicit through millennia of cultural and biological evolution and through universal early childhood experiences. Before machines can share knowledge as flexibly as people do, those prerequisites need to be recapitulated somehow in explicit, computable forms.

For the past decade, researchers at the CYC project in Austin, Texas, have been hard at work doing exactly that. Originally, the group examined snippets of news articles, novels, advertisements and the like, and for each sentence asked: "What did the writer assume that the reader already knew?" It is that prerequisite knowledge, not the content of the text, that had to be codified. This process has led the group to represent 100,000 discrete concepts and about one million pieces of common-sense knowledge about them.

Many of these entities—for example, "BodyOfWater"—do not correspond to a single English word. Conversely, an innocuous word such as "in" turns out to have two dozen meanings, each corresponding to a distinct concept. The way in which you, the reader, are in a room is different from the way the air is in the room, the way the carpet is in the room, the way the paint on the walls is in the room and the way a letter in a desk drawer is in the room. Each way that something can be "in" a place has different implications—the letter can be removed from the room, for instance, whereas the air cannot. Neither the air nor the letter, however, is visible at first glance to someone entering the room.

## What Everyone Knows

Most of these pieces of knowledge turned out not to be facts from an almanac or definitions from a dictionary but rather common observations and widely held beliefs. CYC had to be taught how people eat soup, that children are sometimes frightened by animals and that police in most countries are armed.

To make matters even more complicated, many of the observations we incorporated into CYC's knowledge base contradict one another. By the time a knowledge-based program grows to contain more than 10,000 rules—1 percent of CYC's size—it becomes difficult to add new knowledge without interfering with something already present. We overcame this hurdle by partitioning the knowledge base into hundreds of separate microtheories, or contexts. Like the individual plates in a suit of armor, each context is fairly rigid and consistent, but articulations between them permit apparent contradictions among contexts. CYC knows that Dracula was a vampire, but at the same time it knows that vampires do not exist.

Fictional contexts (such as the one for Bram Stoker's novel) are important because they allow CYC to understand metaphors and use analogies to solve problems. Multiple contexts are also useful for reasoning at different levels of detail, for capturing the beliefs of different age groups, nationalities or historic epochs, and for describing different programs, each of which makes its own assumptions about the situation in which it will be used. We can even use all the brittle idiot savants from past generations of AI by wrapping each one in a context that describes when and how to use it appropriately.

The breadth of CYC's knowledge is evident even in a simple data retrieval application we built in 1994, which fetches images whose descriptions match the criteria a user selects. In response to a request for pictures containing seated people, CYC was able to locate this caption: "There are some cars.

They are on a street. There are some trees on the side of the street. They are shedding their leaves. Some of them are yellow taxicabs. The New York City skyline is in the background. It is sunny." The program then used its formalized common sense about cars—they have a driver's seat, and cars in motion are generally being driven—to infer that there was a good chance the image was relevant. Similarly, CYC could parse the request "Show me happy people" and deliver a picture whose caption reads, "A man watching his daughter learn to walk." None of the words are synonymous or even closely related, but a little common sense makes it easy to find the connection.

## Ready for Takeoff

CYC is far from complete, but it is approaching the level at which it can serve as the seed from which a base of shared knowledge can grow. Programs that understand natural languages will employ the existing knowledge base to comprehend a wide variety of texts laden with ambiguity and metaphor; information drawn from CYC's readings will augment its concepts and thus enable further extensions. CYC will also learn by discovery, forming plausible hypotheses about the world and testing them. One of the provocative analogies it noticed and explored a few years ago was that between a country and a family. Like people, CYC will learn at the fringes of what it already knows, and so its capacity for education will depend strongly on its existing knowledge.

During the coming decade, researchers will flesh out CYC's base of shared knowledge by both manual and automated means. They will also begin to build applications, embedding common sense in familiar sorts of software appliances, such as spreadsheets, databases, document preparation systems and on-line search agents.

Word processors will check content, not just spelling and grammar; if you promise your readers to discuss an issue later in a document but fail to do so, a warning may appear on your screen. Spreadsheets will highlight entries that are technically permissible but violate common sense. Document retrieval programs will understand enough of the content of what they are searching—and of your queries—to find the texts you are looking for regardless of whether they contain the words you specify.

These kinds of programs will act in concert with existing trends in computer hardware and networks to make computer-based services ever less expensive and more ubiquitous, to build steadily better user models and agent software and to immerse the user deeper in virtual environments. The goal of a general artificial intelligence is in sight, and the 21st-century world will be radically changed as a result. The late Allen Newell, one of the field's founders, likened the coming era to the land of Faery: inanimate objects such as appliances conversing with you, not to mention conversing and coordinating with one another. Unlike the creatures of most fairy stories, though, they will generally be plotting to do people good, not ill.

*Classical artificial intelligence requires enormous computing power. One way to cut back on computing cycles, and therefore speed up the learning curve, is to employ a methodology called fuzzy logic. By finding commonalities within statements and employing a "then/if" sequence, machines can, in a fashion, be programmed to think for themselves.*

# Fuzzy Logic

## Bart Kosko

Computers do not reason as brains do. Computers "reason" when they manipulate precise facts that have been reduced to strings of zeros and ones and statements that are either true or false. The human brain can reason with vague assertions or claims that involve uncertainties or value judgments: "The air is cool," or "That speed is fast" or "She is young." Unlike computers, humans have common sense that enables them to reason in a world where things are only partially true.

Fuzzy logic is a branch of machine intelligence that helps computers paint gray, commonsense pictures of an uncertain world. Logicians in the 1920s first broached its key concept: everything is a matter of degree.

Fuzzy logic manipulates such vague concepts as "warm" or "still dirty" and so helps engineers to build air conditioners, washing machines and other devices that judge how fast they should operate or shift from one setting to another even when the criteria for making those changes are hard to define. When mathematicians lack specific algorithms that dictate how a system should respond to inputs, fuzzy logic can control or describe the system by using "commonsense" rules that refer to indefinite quantities. No known mathematical model can back up a truck-and-trailer rig from a parking lot to a loading dock when the vehicle starts from a random spot. Both humans and fuzzy systems can perform this nonlinear guidance task by using practical but imprecise rules such as "If the trailer turns a little to the left, then turn it a little to the right." Fuzzy systems often glean their rules from experts. When no expert gives the rules, adaptive fuzzy systems learn the rules by observing how people regulate real systems.

A recent wave of commercial fuzzy products, most of them from Japan, has popularized fuzzy logic. In 1980 the contracting firm of F.L. Smidth & Company in Copenhagen first used a fuzzy system to oversee the operation of a cement kiln. In 1988 Hitachi turned over control of a subway in Sendai, Japan, to a fuzzy system. Since then, Japanese companies have used fuzzy logic to direct hundreds of household appliances and electronics products. The Ministry of International Trade and Industry estimates that in 1992 Japan produced about $2 billion worth of fuzzy products. U.S. and European companies still lag far behind.

Applications for fuzzy logic extend beyond control systems. Recent theorems show that in principle fuzzy logic can be used to model any continuous system, be it based in engineering or physics or biology or economics. Investigators in many fields may find that fuzzy, common-sense models are more useful or accurate than are standard mathematical ones.

At the heart of the difference between classical and fuzzy logic is something Aristotle called the law of the excluded middle. In standard set theory, an object either does or does not belong to a set. There is no middle ground: the number five belongs fully to the set of odd numbers and not at all to the set of even numbers. In such bivalent sets, an object cannot belong to both a set and its complement set or to neither of the sets. This principle preserves the structure of logic and avoids the contradiction of an object that both is and is not a thing at the same time.

Sets that are fuzzy, or multivalent, break the law of the excluded middle—to some degree. Items belong only partially to a fuzzy set. They may also belong to more than one set. Even to just one individual, the air may feel cool, just right and warm to varying degrees. Whereas the boundaries of standard sets are exact, those of fuzzy sets are curved or taper off, and this curvature creates partial contradictions. The air can be 20 percent cool—and at the same time, 80 percent not cool.

Fuzzy degrees are not the same as probability percentages, a point that has eluded some critics of the field. Probabilities measure whether something will occur or not. Fuzziness measures the degree to which something occurs or some condition exists. The statement "There is a 30 percent chance the weather will be cool" conveys the probability of cool weather. But "The morning feels 30 percent cool" means that the air feels cool to some extent—and at the same time, just right and warm to varying extents.

The only constraint on fuzzy logic is that an object's degrees of membership in complementary groups must sum to unity. If the air seems 20 percent cool, it must also be 80 percent not cool. In this way, fuzzy logic just skirts the bivalent contradiction—that something is 100 percent cool and 100 percent not cool—that would destroy formal logic. The law of the excluded middle holds merely as a special case in fuzzy logic, namely when an object belongs 100 percent to one group.

The modern study of fuzzy logic and partial contradictions had its origins early in this century, when Bertrand Russell found the ancient Greek paradox at the core of modern set theory and logic. According to the old riddle, a Cretan asserts that all Cretans lie. So, is he lying? If he lies, then he tells the truth and does not lie. If he does not lie, then he tells the truth and so lies. Both cases lead to a contradiction because the statement is both true and false. Russell found the same paradox in set theory. The set of all sets is a set, and so it is a member of itself. Yet the set of all apples is not a member of itself because its members are apples and not sets. Perceiving the underlying contradiction, Russell then asked, "Is the set of all sets that are not members of themselves a member of itself?" If it is, it isn't; if it isn't, it is.

Faced with such a conundrum, classical logic surrenders. But fuzzy logic says that the answer is half true and half false, a 50–50 divide. Fifty percent of the Cretan's statements are true, and 50 percent are false. The Cretan lies 50 percent of the time and does not lie the other half. When membership is less than total, a bivalent system might simplify the problem by rounding it down to zero or up to 100 percent. Yet 50 percent does not round up or down.

In the 1920s, independent of Russell, the Polish logician Jan Wukasiewicz worked out the principles of multivalued logic, in which statements can take on fractional truth values between the ones and zeros of binary logic. In a 1937 article in *Philosophy of Science,* quantum philosopher Max Black applied multivalued logic to lists, or sets of objects, and in so doing drew the first fuzzy set curves. Following Russell's lead, Black called the sets "vague."

Almost 30 years later Lotfi A. Zadeh, then chair of the electrical engineering department at the University of California at Berkeley, published "Fuzzy Sets," a landmark paper that gave the field its name. Zadeh applied Wukasiewicz's logic to every object in a set and worked out a complete algebra for fuzzy sets. Even so, fuzzy sets were not put to use until the mid-1970s, when Ebrahim H. Mamdani of Queen Mary College in London designed a fuzzy controller for a steam engine. Since then, the term "fuzzy logic" has come to mean any mathematical or computer system that reasons with fuzzy sets.

*    *    *

Fuzzy logic is based on rules of the form "if . . . then" that convert inputs to outputs—one fuzzy set into another. The controller of a car's air conditioner might include rules such as "If the temperature is cool, then set the motor speed on slow" and "If the temperature is just right, then set the motor speed on medium." The temperatures (cool, just right) and the motor speeds (slow, medium) name fuzzy sets rather than specific values.

To build a fuzzy system, an engineer might begin with a set of fuzzy rules from an expert. An engineer might define the degrees of membership in various fuzzy input and output sets with sets of curves. The relation between the input and output sets could then be plotted. Given the rule "If the air feels cool, then set the motor to slow," the inputs (temperature) would be listed along one axis of a graph and the outputs (motor speed) along a second axis. The product of these fuzzy sets forms a fuzzy patch, an area that represents the set of all associations that the rule forms between those inputs and outputs.

The size of the patch reflects the rule's vagueness or uncertainty. The more precise the fuzzy set, the smaller it becomes. If "cool" is precisely 68 degrees Fahrenheit, the fuzzy set collapses to a spike. If both the cool and slow fuzzy sets are spikes, the rule patch is a point.

The rules of a fuzzy system define a set of overlapping patches that relate a full range of inputs to a full range of outputs. In that sense, the fuzzy system approximates some mathematical function or equation of cause and effect. These functions might be laws that tell a microprocessor how to adjust the power of an air conditioner or the speed of a washing machine in response to some fresh measurement.

Fuzzy systems can approximate any continuous math function. I proved this uniform convergence theorem by showing that enough small fuzzy patches can sufficiently cover the graph of any function or input/output relation. The theorem also shows that we can pick in advance the maximum error of the approximation and be sure there is a finite number of fuzzy rules that achieve it. A fuzzy system reasons, or infers, based on its rule patches. Two or more rules convert any incoming number into some result because the patches overlap. When data trigger the rules, overlapping patches fire in parallel—but only to some degree.

Imagine a fuzzy air conditioner that relies on five rules and thus five patches to match temperatures to motor speeds. The temperature sets (cold, cool, just right, warm and hot) cover all the possible fuzzy inputs. The motor speed sets (very slow, slow, medium, fast and very fast) describe all the fuzzy outputs. A temperature of, say, 68 degrees F might be 20 percent cool (80 percent not cool) and 70 percent just right (30 percent not just right). At the same time, the air is also 0 percent cold, warm and hot. The "if cool" and "if just right" rules would fire and invoke both the slow and medium motor speeds.

The two rules contribute proportionally to the final motor speed. Because the temperature was 20 percent cool, the curve describing the slow engine speed must shrink to 20 percent of its height. The "medium" curve must shrink to 70 percent. Summing those two reduced curves produces the final curve for the fuzzy output set.

In its fuzzy form, such an output curve does not assist controllers that act on binary instructions. So the final step is a process of defuzzification, in which the fuzzy output curve is turned into a single numerical value. The most common technique is to compute the center of mass, or centroid, of the area under the curve. In this instance, the centroid of the fuzzy output curve might correspond to a motor speed of 47 revolutions per minute. Thus, beginning with a quantitative temperature input, the electronic controller can reason from fuzzy temperature and motor speed sets and arrive at an appropriate and precise speed output.

All fuzzy systems reason with this fire-and-sum technique—or something close to it. As systems become more complex, the antecedents of the rules may include any number of terms conjoined by "and" or disjoined by "or." An advanced fuzzy air conditioner might use a rule that says, "If the air is cool and the humidity is high, then set the motor to medium."

Fuzzy products use both microprocessors that run fuzzy inference algorithms and sensors that measure changing input conditions. Fuzzy chips are microprocessors designed to store and process fuzzy rules. In 1985 Masaki Togai and Hiroyuki Watanabe, then working at AT&T Bell Laboratories, built the first digital fuzzy chip. It processed 16 simple rules in 12.5 microseconds, a rate of 0.08 million fuzzy logical inferences per second. Togai InfraLogic, Inc., now offers chips based on Fuzzy Computational Acceleration hardware that processes up to two million rules per second. Most microprocessor firms currently have fuzzy chip research projects. Fuzzy products largely rely on standard microprocessors that engineers have programmed with a few lines of fuzzy inference code. Although the market for dedicated fuzzy chips is still tiny, the value of microprocessors that include fuzzy logic already exceeds $1 billion.

The most famous fuzzy application is the subway car controller used in Sendai, which has outperformed both human operators and conventional automated controllers. Conventional controllers start or stop a train by reacting to position markers that show how far the vehicle is from a station. Because the controllers are rigidly programmed, the ride may be jerky: the automated controller will apply the same brake pressure when a train is, say, 100 meters from a station, even if the train is going uphill or downhill.

In the mid-1980s engineers from Hitachi used fuzzy rules to accelerate, slow and brake the subway trains more smoothly than could a deft human operator. The rules encompassed a broad range of variables about the ongoing performance of the train, such as how frequently and by how much its speed changed and how close the actual speed was to the maximum speed. In simulated tests the fuzzy controller beat an automated version on measures of riders' comfort, shortened riding times and even achieved a 10 percent reduction in the train's energy consumption. Today the fuzzy system runs the Sendai subway during peak hours and runs some Tokyo trains as well. Humans operate the subway during nonpeak hours to keep up their skills.

Companies in Japan and Korea are building an array of fuzzy consumer goods that offer more precise control than do conventional ones. Fuzzy washing machines adjust the wash cycle to every set of clothes, changing strategies as the clothes become clean. A fuzzy washing machine gives a finer wash than a "dumb" machine with fixed commands. In the simplest of these machines, an optical sensor measures the murk or clarity of the wash water, and the controller estimates how long it would take a stain to dissolve or saturate in the wash water. Some machines use a load sensor to trigger changes in the agitation rate or water temperature. Others shoot bubbles into the wash to help dissolve dirt and detergent. A washing machine may use as few as 10 fuzzy rules to determine a wide variety of washing strategies.

In cameras and camcorders, fuzzy logic links image data to various lens settings. One of the first fuzzy camcorders, the Canon hand-held H800, which was introduced in 1990, adjusts the autofocus based on 13 fuzzy rules. Sensors measure the clarity of images in six areas. The rules take up about a kilobyte of memory and convert the sensor data to new lens settings.

Matsushita relies on more rules to cancel the image jitter that a shaking hand causes in its small Panasonic camcorders. The fuzzy rules infer where the image will shift. The rules heed local and global changes in the image and then compensate for them. In contrast, camcorder controllers based on mathematical models can compensate for no more than a few types of image jitter.

Systems with fuzzy controllers are often more energy efficient because they calculate more precisely how much power is needed to get a job done. Mitsubishi and Korea's Samsung report that their fuzzy vacuum cleaners achieve more than 40 percent energy savings over nonfuzzy designs. The fuzzy systems use infrared light-emitting diodes to measure changes in dust flow and so to judge if a floor is bare. A four-bit microprocessor measures the dust flow to calculate the appropriate suction power and other vacuum settings.

Automobiles also benefit from fuzzy logic. General Motors uses a fuzzy transmission in its Saturn. Nissan has patented a fuzzy antiskid braking system, fuzzy transmission system and fuzzy fuel injector. One set of fuzzy rules in an on-board microprocessor adjusts the fuel flow. Sensors measure the throttle setting, manifold pressure, radiator water temperature and the engine's revolutions per minute. A second set of fuzzy rules times the engine ignition based on the revolutions per minute, water temperature and oxygen concentration.

One of the most complex fuzzy systems is a model helicopter, designed by Michio Sugeno of the Tokyo Institute of Technology. Four elements of the craft—the elevator, aileron, throttle and rudder—respond to 13 fuzzy voice commands, such as "up," "land" and "hover." The fuzzy controller can make the craft hover in place, a difficult task even for human pilots.

A few fuzzy systems manage information rather than devices. With fuzzy logic rules, the Japanese conglomerate Omron oversees five medical data bases in a health management system for large firms. The fuzzy systems use 500 rules to diagnose the health of some 10,000 patients and to draw up personalized plans to help them prevent disease, stay fit and reduce stress. Other companies, including Hitachi and Yamaichi Securities, have built trading programs for bonds or stock funds that use fuzzy rules to react to changes in economic data.

\*   \*   \*

The Achilles' heel of a fuzzy system is its rules. Almost all the fuzzy consumer products now on the market rely on rules supplied by an expert. Engineers then engage in a lengthy process of tuning those rules and the fuzzy sets. To automate this process, some engineers are building adaptive fuzzy systems that use neural networks or other statistical tools to refine or even form those initial rules.

Neural networks are collections of "neurons" and "synapses" that change their values in response to inputs from surrounding neurons and synapses. The neural net acts like a computer because it maps inputs to outputs. The neurons and synapses may be silicon components or equations in software that simulate their behavior. A neuron adds up all the incoming signals from other neurons and then emits its own response in the form of a number. Signals travel across the synapses, which have numerical values that weight the flow of neuronal signals. When new input data fire a network's neurons, the synaptic values can change slightly. A neural net "learns" when it changes the value of its synapses.

Depending on the available data, networks can learn patterns with or without supervision. A supervised net learns by trial and error, guided by a teacher. A human may point out when the network has erred—when it has emitted a response that differs from the desired output. The teacher will correct the responses to sample data until the network responds correctly to every input.

Supervised networks tune the rules of a fuzzy system as if they were synapses. The user provides the first set of rules, which the neural net refines by running through hundreds of thousands of inputs, slightly varying the fuzzy sets each time to see how well the system performs. The net tends to keep the changes that improve performance and to ignore the others.

A handful of products in Japan now use supervised neural learning to tune the fuzzy rules that control their operation. Among them are Sanyo's microwave oven and several companies' washing machines. Sharp employs this technique to modify the rules of its fuzzy refrigerator so that the device learns how often its hungry patron is likely to open the door and adjusts the cooling cycle accordingly. So far the neural net must learn "off-line" in the laboratory, from small samples of behavior by average customers. In time, researchers at such groups as Japan's Laboratory for International Fuzzy Engineering and the Fuzzy Logic Systems Institute hope to build fuzzy systems that will adapt to the needs of each consumer.

Supervised networks do have drawbacks. Tuning such systems can take hours or days of computer time because networks may converge on an inappropriate solution or rule or may fail to converge at all. Neural researchers have proposed hundreds of schemes to alleviate this problem, but none has removed it. Even after a lengthy tuning session, the final rules may not be much better than the first set.

Rather than relying on an expert to supply a training set of data and correct a network in the process of learning, unsupervised neural networks learn simply by observing an expert's decisions. In this way, an adaptive fuzzy system can learn to spot rule patterns in the incoming data. Broad rule patches form quickly, based on a few inputs. Those patches are refined over time.

Unsupervised neural networks blindly cluster data into groups, the members of which resemble one another. There may be no given right or wrong response or way to organize the data. The algorithms are simpler, and, at least in theory, the network need run through the data just once. (In some cases, when data are sparse, the neural net must cycle through them repeatedly.) Unsupervised learning is thus much faster than supervised learning. With numerical inputs and outputs supplied by an expert or a physical process or even an algorithm, an unsupervised neural network can find the first set of rules for a fuzzy system. The quality of the rules depends on the quality of the data and therefore on the skills of the expert who generates the data. At this point, there are fewer unsupervised than supervised adaptive fuzzy systems. Because unsupervised networks are best used to create rules and supervised networks are better at refining them, hybrid adaptive fuzzy systems include both.

Most fuzzy systems have been control systems with few variables. That trend happened because most of the first fuzzy logic engineers were control theorists and because a control loop regulates most consumer products. The challenge for the next generation of fuzzy research will be tackling large-scale, nonlinear systems with many variables. These problems can arise when people try to supervise manufacturing plants or schedule airline flights or model the economy. No experts may be able to describe such systems. Common sense may falter or not apply. The neural nets that must learn the rules for modeling these hard problems may have little or no data to go on.

A further problem is that, like any other mathematical or computer model, fuzzy logic falls prey to the "curse of dimensionality": the number of fuzzy rules tends to grow exponentially as the number of system variables increases. Fuzzy systems must contend with a trade-off. Large rule patches mean the system is more manageable but also less precise. Even with that trade-off, fuzzy logic can often better model the vagueness of the world than can the black-and-white concepts of set theory. For that reason, fuzzy logic systems may well find their way into an ever growing number of computers, home appliances and theoretical models. The next century may be fuzzier than we think.

*Could a computer learn and respond in a humanlike way by patterning the bioneural network of the human body into an operating system? This idea, sparked by AI maverick Marvin Minsky, took root and now is a serious line of inquiry in the field. These inquiries have expanded a whole new appreciation of the miracle of human perception.*

# How Neural Networks Learn from Experience

## Geoffrey E. Hinton

T he brain is a remarkable computer. It interprets imprecise information from the senses at an incredibly rapid rate. It discerns a whisper in a noisy room, a face in a dimly lit alley and a hidden agenda in a political statement. Most impressive of all, the brain learns—without any explicit instructions—to create the internal representations that make these skills possible.

Much is still unknown about how the brain trains itself to process information, so theories abound. To test these hypotheses, my colleagues and I have attempted to mimic the brain's learning processes by creating networks of artificial neurons. We construct these neural networks by first trying to deduce the essential features of neurons and their interconnections. We then typically program a computer to simulate these features.

Because our knowledge of neurons is incomplete and our computing power is limited, our models are necessarily gross idealizations of real networks of neurons. Naturally, we enthusiastically debate what features are most essential in simulating neurons. By testing these features in artificial neural networks, we have been successful at ruling out all kinds of theories about how the brain processes information. The models are also beginning to reveal how the brain may accomplish its remarkable feats of learning.

In the human brain, a typical neuron collects signals from others through a host of fine structures called dendrites. The neuron sends out spikes of electrical activity through a long, thin strand known as an axon, which splits into thousands of branches. At the end of each branch, a structure called a synapse converts the activity from the axon into electrical effects that inhibit or excite activity in the connected neurons. When a neuron receives excitatory input that is sufficiently large compared with its inhibitory input, it sends a spike of electrical activity down its axon. Learning occurs by changing the effectiveness of the synapses so that the influence of one neuron on another changes.

Artificial neural networks are typically composed of interconnected "units," which serve as model neurons. The function of the synapse is modeled by a modifiable weight, which is associated with each connection. Most artificial networks do not reflect the detailed geometry of the dendrites and axons, and they express the electrical output of a neuron as a single number that represents the rate of firing—its activity.

The network of neurons in the brain provides people with the ability to assimilate information. Will simulations of such networks reveal the underlying mechanisms of learning?

Each unit converts the pattern of incoming activities that it receives into a single outgoing activity that it broadcasts to other units. It performs this conversion in two stages. First, it multiplies each incoming activity by the weight on the connection and adds together all these weighted inputs to get a quantity called the total input. Second, a unit uses an input-output function that transforms the total input into the outgoing activity.

The behavior of an artificial neural network depends on both the weights and the input-output function that is specified for the units. This function typically falls into one of three categories: linear, threshold or sigmoid. For linear units, the output activity is proportional to the total weighted input. For threshold units, the output is set at one of two levels, depending on whether the total input is greater than or less than some threshold value. For sigmoid units, the output varies continuously but not linearly as the input changes. Sigmoid units bear a greater resemblance to real neurons than do linear or threshold units, but all three must be considered rough approximations.

To make a neural network that performs some specific task, we must choose how the units are connected to one another, and we must set the weights on the connections appropriately. The connections determine whether it is possible for one unit to influence another. The weights specify the strength of the influence.

The commonest type of artificial neural network consists of three groups, or layers, of units: a layer of input units is connected to a layer of "hidden" units, which is connected to a layer of output units. The activity of the input units represents the raw information that is fed into the network. The activity of each hidden unit is determined by the activities of the input units and the weights on the connections between the input and hidden units. Similarly, the behavior of the output units depends on the activity of the hidden units and the weights between the hidden and output units.

This simple type of network is interesting because the hidden units are free to construct their own representations of the input. The weights between the input and hidden units determine when each hidden unit is active, and so by modifying these weights, a hidden unit can choose what it represents.

We can teach a three-layer network to perform a particular task by using the following procedure. First, we present the network with training examples, which consist of a pattern of activities for the input units together with the desired pattern of activities for the output units. We then determine how closely the actual output of the network matches the desired output. Next we change the weight of each connection so that the network produces a better approximation of the desired output.

For example, suppose we want a network to recognize handwritten digits. We might use an array of, say, 256 sensors, each recording the presence or absence of ink in a small area of a single digit. The network would therefore need 256 input units (one for each sensor), 10 output units (one for each kind of digit) and a number of hidden units. For each kind of digit recorded by the sensors, the network should produce high activity in the appropriate output unit and low activity in the other output units.

To train the network, we present an image of a digit and compare the actual activity of the 10 output units with the desired activity. We then calculate the error, which is defined as the square of the difference between the actual and the desired activities. Next we change the weight of each connection so as to reduce the error. We repeat this training process for many different images of each kind of digit until the network classifies every image correctly.

To implement this procedure, we need to change each weight by an amount that is proportional to the rate at which the error changes as the weight is changed. This quantity— called the error derivative for the weight, or simply the EW—is tricky to compute efficiently. One way to calculate the EW is to perturb a weight slightly and observe how the error changes. But that method is inefficient because it requires a separate perturbation for each of the many weights.

Around 1974 Paul J. Werbos invented a much more efficient procedure for calculating the EW while he was working toward a doctorate at Harvard University. The procedure, now known as the back-propagation algorithm, has become one of the more important tools for training neural networks.

The back-propagation algorithm is easiest to understand if all the units in the network are linear. The algorithm computes each EW by first computing the EA, the rate at which the error changes as the activity level of a unit is changed. For output units, the EA is simply the difference between the actual and the desired output. To compute the EA for a hidden unit in the layer just before the output layer, we first identify all the weights between that hidden unit and the output units to which it is connected. We then multiply those weights by the EAs of those output units and add the products. This sum equals the EA for the chosen hidden unit. After calculating all the EAs in the hidden layer just before the output layer, we can compute in like fashion the EAs for other layers, moving from layer to layer in a direction opposite to the way activities propagate through the network. This is what gives back propagation its name. Once the EA has been computed for a unit, it is straightforward to compute the EW for each incoming connection of the unit. The EW is the product of the EA and the activity through the incoming connection.

For nonlinear units, the back-propagation algorithm includes an extra step. Before back-propagating, the EA must be converted into the EI, the rate at which the error changes as the total input received by a unit is changed.

The back-propagation algorithm was largely ignored for years after its invention, probably because its usefulness was not fully appreciated. In the early 1980s David E. Rumelhart, then at the University of California at San Diego, and David B. Parker, then at Stanford University, independently rediscovered the algorithm. In 1986 Rumelhart, Ronald J. Williams and I popularized the algorithm by demonstrating that it could teach the hidden units to produce interesting representations of complex input patterns.

The back-propagation algorithm has proved surprisingly

good at training networks with multiple layers to perform a wide variety of tasks. It is most useful in situations in which the relation between input and output is nonlinear and training data are abundant. By applying the algorithm, researchers have produced neural networks that recognize handwritten digits, predict currency exchange rates and maximize the yields of chemical processes. They have even used the algorithm to train networks that identify precancerous cells in Pap smears and that adjust the mirror of a telescope so as to cancel out atmospheric distortions.

Within the field of neuroscience, Richard Andersen and David Zipser showed that the back-propagation algorithm is a useful tool for explaining the function of some neurons in the brain's cortex. They trained a neural network to respond to visual stimuli using back propagation. They then found that the responses of the hidden units were remarkably similar to those of real neurons responsible for converting visual information from the retina into a form suitable for deeper visual areas of the brain.

Yet back propagation has had a rather mixed reception as a theory of how biological neurons learn. On the one hand; the back-propagation algorithm has made a valuable contribution at an abstract level. The algorithm is quite good at creating sensible representations in the hidden units. As a result, researchers gained confidence in learning procedures in which weights are gradually adjusted to reduce errors. Previously, many workers had assumed that such methods would be hopeless because they would inevitably lead to locally optimal but globally terrible solutions. For example, a digit-recognition network might consistently home in on a set of weights that makes the network confuse ones and sevens even though an ideal set of weights exists that would allow the network to discriminate between the digits. This fear supported a widespread belief that a learning procedure was interesting only if it were guaranteed to converge eventually on the globally optimal solution.

Back propagation showed that for many tasks global convergence was not necessary to achieve good performance.

On the other hand, back propagation seems biologically implausible. The most obvious difficulty is that information must travel through the same connections in the reverse direction, from one layer to the previous layer. Clearly, this does not happen in real neurons. But this objection is actually rather superficial. The brain has many pathways from later layers back to earlier ones, and it could use these pathways in many ways to convey the information required for learning.

A more important problem is the speed of the back-propagation algorithm. Here the central issue is how the time required to learn increases as the network gets larger. The time taken to calculate the error derivatives for the weights on a given training example is proportional to the size of the network because the amount of computation is proportional to the number of weights. But bigger networks typically require more training examples, and they must update the weights more times. Hence, the learning time grows much faster than does the size of the network.

The most serious objection to back propagation as a model of real learning is that it requires a teacher to supply the desired output for each training example. In contrast, people learn most things without the help of a teacher. Nobody presents us with a detailed description of the internal representations of the world that we must learn to extract from our sensory input. We learn to understand sentences or visual scenes without any direct instructions.

How can a network learn appropriate internal representations if it starts with no knowledge and no teacher? If a network is presented with a large set of patterns but is given no information about what to do with them, it apparently does not have a well-defined problem to solve. Nevertheless, researchers have developed several general-purpose, unsupervised procedures that can adjust the weights in the network appropriately.

All these procedures share two characteristics: they appeal, implicitly or explicitly, to some notion of the quality of a representation, and they work by changing the weights to improve the quality of the representation extracted by the hidden units.

In general, a good representation is one that can be described very economically but nonetheless contains enough information to allow a close approximation of the raw input to be reconstructed. For example, consider an image consisting of several ellipses. Suppose a device translates the image into an array of a million tiny squares, each of which is either light or dark. The image could be represented simply by the positions of the dark squares. But other, more efficient representations are also possible. Ellipses differ in only five ways: orientation, vertical position, horizontal position, length and width. The image can therefore be described using only five parameters per ellipse.

Although describing an ellipse by five parameters requires more bits than describing a single dark square by two coordinates, we get an overall savings because far fewer parameters than coordinates are needed. Furthermore, we do not lose any information by describing the ellipses in terms of their parameters: given the parameters of the ellipse, we could reconstruct the original image if we so desired.

Almost all the unsupervised learning procedures can be viewed as methods of minimizing the sum of two terms, a code cost and a reconstruction cost. The code cost is the number of bits required to describe the activities of the hidden units. The reconstruction cost is the number of bits required to describe the misfit between the raw input and the best approximation to it that could be reconstructed from the activities of the hidden units. The reconstruction cost is proportional to the squared difference between the raw input and its reconstruction.

Two simple methods for discovering economical codes allow fairly accurate reconstruction of the input principal-components learning and competitive learning. In both approaches, we first decide how economical the code should be and then modify the weights in the network to minimize the reconstruction error.

A principal-components learning strategy is based on the idea that if the activities of pairs of input units are correlated in some way, it is a waste of bits to describe each input activity separately. A more efficient approach is to extract and describe the principal components—that is, the components of variation shared by many input units. If we wish to discover, say, 10 of the principal components, then we need only a single layer of 10 hidden units.

Because such networks represent the input using only a small number of components, the code cost is low. And because the input can be reconstructed quite well from the principal components, the reconstruction cost is small.

One way to train this type of network is to force it to reconstruct an approximation to the input on a set of output units. Then back propagation can be used to minimize the difference between the actual output and the desired output. This process resembles supervised learning, but because the desired output is exactly the same as the input, no teacher is required.

Many researchers, including Ralph Linsker and Erkki Oja have discovered alternative algorithms for learning principal components. These algorithms are more biologically plausible because they do not require output units or back propagation. Instead they use the correlation between the activity of a hidden unit and the activity of an input unit to determine the change in the weight.

When a neural network uses principal-components learning, a small number of hidden units cooperate in representing the input pattern. In contrast, in competitive learning, a large number of hidden units compete so that a single hidden unit is used to represent any particular input pattern. The selected hidden unit is the one whose incoming weights are most similar to the input pattern.

Now suppose we had to reconstruct the input pattern solely from our knowledge of which hidden unit was chosen. Our best bet would be to copy the pattern of incoming weights of the chosen hidden unit. To minimize the reconstruction error, we should move the pattern of weights of the winning hidden unit even closer to the input pattern. This is what competitive learning does. If the network is presented with training data that can be grouped into clusters of similar input patterns, each hidden unit learns to represent a different cluster, and its incoming weights converge on the center of the cluster.

Like the principal-components algorithm, competitive learning minimizes the reconstruction cost while keeping the code cost low. We can afford to use many hidden units because even with a million units it takes only 20 bits to say which one won.

In the early 1980s Teuvo Kohonen introduced an important modification of the competitive learning algorithm. Kohonen showed how to make physically adjacent hidden units learn to represent similar input patterns. Kohonen's algorithm adapts not only the weights of the winning hidden unit but also the weights of the winner's neighbors. The algorithm's ability to map similar input patterns to nearby hidden units suggests that a procedure of this type may be what the brain uses to create the topographic maps found in the visual cortex.

Unsupervised learning algorithms can be classified according to the type of representation they create. In principal-components methods, the hidden units cooperate, and the representation of each input pattern is distributed across all of them. In competitive methods, the hidden units compete, and the representation of the input pattern is localized in the single hidden unit that is selected. Until recently, most work on unsupervised learning focused on one or another of these two techniques, probably because they lead to simple rules for changing the weights. But the most interesting and powerful algorithms probably lie somewhere between the extremes of purely distributed and purely localized representations.

Horace B. Barlow has proposed a model in which each hidden unit is rarely active and the representation of each input pattern is distributed across a small number of selected hidden units. He and his co-workers have shown that this type of code can be learned by forcing hidden units to be uncorrelated while also ensuring that the hidden code allows good reconstruction of the input.

Unfortunately, most current methods of minimizing the code cost tend to eliminate all the redundancy among the activities of the hidden units. As a result, the network is very sensitive to the malfunction of a single hidden unit. This feature is uncharacteristic of the brain, which is generally not affected greatly by the loss of a few neurons.

The brain seems to use what are known as population codes, in which information is represented by a whole population of active neurons. That point was beautifully demonstrated in the experiments of David L. Sparks and his co-workers. While investigating how the brain of a monkey instructs its eyes where to move, they found that the required movement is encoded by the activities of a whole population of cells, each of which represents a somewhat different movement. The eye movement that is actually made corresponds to the average of all the movements encoded by the active cells. If some brain cells are anesthetized, the eye moves to the point associated with the average of the remaining active cells. Population codes may be used to encode not only eye movements but also faces, as shown by Malcolm P. Young and Shigeru Yamane at the RIKEN Institute in Japan in recent experiments on the inferior temporal cortex of monkeys.

For both eye movements and faces, the brain must represent entities that vary along many different dimensions. In the case of an eye movement, there are just two dimensions, but for something like a face, there are dimensions such as happiness, hairiness or familiarity, as well as spatial parameters such as position, size and orientation. If we associate with each face-sensitive cell the parameters of the face that make it most active, we can average these parameters over a population of active cells to discover the parameters of the face being represented by that population code. In abstract terms, each face cell represents a particular point in a multidimensional space of possible faces, and any face can then be represented by activating all the cells that encode very similar faces, so that a bump of activity appears in the multidimensional space of possible faces.

Population coding is attractive because it works even if some of the neurons are damaged. It can do so because the loss of a random subset of neurons has little effect on the population average. The same reasoning applies if some neurons are overlooked when the system is in a hurry. Neurons communicate by sending discrete spikes called action potentials, and in a very short time interval many of the "active" neurons may not have time to send a spike. Nevertheless, even in such a short interval, a population code in one part of the brain can still give rise to an approximately correct population code in another part of the brain.

At first sight, the redundancy in population codes seems incompatible with the idea of constructing internal representations that minimize the code cost. Fortunately, we can overcome this difficulty by using a less direct measure of code cost. If the activity that encodes a particular entity is a smooth bump in which activity falls off in a standard way as we move away from the center, we can describe the bump of activity completely merely by specifying its center. So a fairer measure of code cost is the cost of describing the center of the bump of activity plus the cost of describing how the actual activities of the units depart from the desired smooth bump of activity.

Using this measure of the code cost, we find that population codes are a convenient way of extracting a hierarchy of progressively more efficient encodings of the sensory input. This point is best illustrated by a simple example. Consider a neural network that is presented with an image of a face. Suppose the network already contains one set of units dedicated to representing noses, another set for mouths and another set for eyes. When it is shown a particular face, there will be one bump of activity in the nose units, one in the mouth units and two in the eye units. The location of each of these activity bumps represents the spatial parameters of the feature encoded by the bump. Describing the four activity bumps is cheaper than describing the raw image, but it would obviously be cheaper still to describe a single bump of activity in a set of face units, assuming of course that the nose, mouth and eyes are in the correct spatial relations to form a face.

This raises an interesting issue: How can the network check that the parts are correctly related to one another to make a face? Some time ago Dana H. Ballard introduced a clever technique for solving this type of problem that works nicely with population codes.

If we know the position, size and orientation of a nose, we can predict the position, size and orientation of the face to which it belongs because the spatial relation between noses and faces is roughly fixed. We therefore set the weights in the neural network so that a bump of activity in the nose units tries to cause an appropriately related bump of activity in the face units. But we also set the thresholds of the face units so that the nose units alone are insufficient to activate the face units. If, however, the bump of activity in the mouth units also tries to cause a bump in the same place in the face units, then the thresholds can be overcome. In effect, we have checked that the nose and mouth are correctly related to each other by checking that they both predict the same spatial parameters for the whole face.

This method of checking spatial relations is intriguing because it makes use of the kind of redundancy between dif-

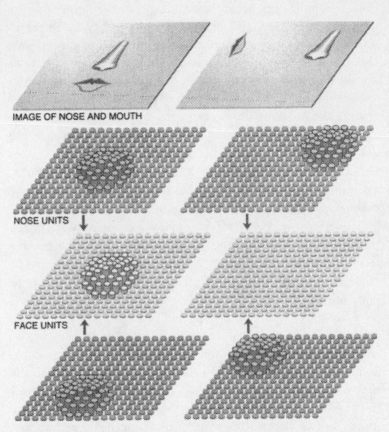

IMAGE OF NOSE AND MOUTH

NOSE UNITS

FACE UNITS

MOUTH UNITS

How can a neural network recognize a face? If the network knows the general spatial relationship between the eyes, nose, and mouth in relation to the face, the units add up to a recognition. In the case where the mouth and nose are out of phase with programmed face parameters (column on the right), the machine will not identify the pattern as a face.

ferent parts of an image that unsupervised learning should be good at finding. It therefore seems natural to try to use unsupervised learning to discover hierarchical population codes for extracting complex shapes. In 1986 Eric Saund of M.I.T. demonstrated one method of learning simple population codes for shapes. It seems likely that with a clear definition of the code cost, an unsupervised network will be able to discover more complex hierarchies by trying to minimize the cost of coding the image. Richard Zemel and I are now investigating this possibility.

By using unsupervised learning to extract a hierarchy of successively more economical representations, it should be possible to improve greatly the speed of learning in large multilayer networks. Each layer of the network adapts its incoming weights to make its representation better than the representation in the previous layer, so weights in one layer can be learned without reference to weights in subsequent layers. This strategy eliminates many of the interactions between weights that make back-propagation learning very slow in deep multilayer networks.

All the learning procedures discussed thus far are implemented in neural networks in which activity flows only in the forward direction from input to output even though error derivatives may flow in the backward direction. Another important possibility to consider is networks in which activity flows around closed loops. Such recurrent networks may settle down to stable states, or they may exhibit complex temporal dynamics that can be used to produce sequential behavior. If they settle to stable states, error derivatives can be computed using methods much simpler than back propagation.

Although investigators have devised some powerful learning algorithms that are of great practical value, we still do not know which representations and learning procedures are actually used by the brain. But sooner or later computational studies of learning in artificial neural networks will converge on the methods discovered by evolution. When that happens, a lot of diverse empirical data about the brain will suddenly make sense, and many new applications of artificial neural networks will become feasible.

Speed. *It's one of the key requirements for flexible, useable artificial intelligence. The future of silicon-based microchips looks dim. But the budding field of nanotechnology may provide a speed solution with specific molecules that can be assembled into microcircuitry.*

# Computing with Molecules

## Mark A. Reed and James M. Tour

How fast and powerful can computers become? Will it be possible someday to create artificial "brains" that have intellectual capabilities comparable—or even superior—to those of human beings? The answers to these questions depend to a very great extent on a single factor: how small and dense we can make computer circuits.

Few if any researchers believe that our present technology—semiconductor-based solid-state microelectronics—will lead to circuitry dense and complex enough to give rise to true cognitive abilities. And until recently, none of the technologies proposed as successors to solid-state microelectronics had shown enough promise to rise above the pack. Within the past year, however, scientists have achieved revolutionary advances that may very well radically change the future of computing. And although the road from here to intelligent machines is still rather long and might turn out to have unbridgeable gaps, the fact that there is a potential path at all is something of a triumph.

The recent advances were in molecular-scale electronics, a

field emerging around the premise that it is possible to build individual molecules that can perform functions identical or analogous to those of the transistors, diodes, conductors and other key components of today's microcircuits. After a period of high hopes but few tangible results, several developments over the past few years have raised expectations that this technology may one day provide the building blocks for future generations of ultrasmall, ultradense electronic computer logic. In a remarkable series of demonstrations, chemists, physicists and engineers have shown that individual molecules can conduct and switch electric current and store information.

In July of 1999, in an achievement widely reported in the popular press, researchers from Hewlett-Packard and the University of California at Los Angeles announced that they had built an electronic switch consisting of a layer of several million molecules of an organic substance called rotaxane. By linking a number of switches, the researchers produced a rudimentary version of an AND gate, a device that performs a basic logic operation. With well over a million molecules apiece, the switches are far larger than would be desirable. And they could be switched only one time before becoming inoperable. Nevertheless, their assembly into a logic gate was of fundamental significance.

Within months of that announcement, our groups at Yale and Rice universities published results on a different class of molecules that acted as a reversible switch. And one month later we described a molecule we had created that could change its electrical conductivity by storing electrons on demand, acting as a memory device.

To produce our switch, we inserted regions into the molecules that trapped electrons, but only when the molecules were subjected to certain voltages. Thus, the degree to which the molecules resisted a flow of electrons depended on the voltage applied to them. In fact, by varying the voltage, we could repeatedly change the molecules at will from a conducting to a nonconducting state—which is the basic requirement for an electrical switch. The tiny device actually consisted of a layer of about 1,000 molecules of nitroamine benzenethiol sandwiched between metal contacts.

After creating the switch, we realized that if we could redesign the molecule so that it could retain electrons rather than trapping them briefly, we would have something that could work as a memory element. We went to work on the trapping region of the molecule, modifying it so that its conductivity could be changed repeatedly. The resulting "electron sucker" could retain electrons for nearly 10 minutes—compared with a few milliseconds for conventional silicon-based dynamic random-access memory.

Although the advances were encouraging, the challenges remaining are enormous. Creating individual devices is an essential first step. But before we can build complete, useful circuits we must find a way to secure many millions, if not billions, of molecular devices of various types against some kind of immobile surface and to link them in any manner and into whatever patterns our circuit diagrams dictate. The technology is still too young to say for sure whether this monumental challenge will ever be surmounted.

## The End of the Road Map

Given the magnitude of the challenges ahead, why did researchers and even the mainstream media pay so much attention to the recent advances? The answer has to do with industrial society's dependence on microelectronics—and the limits of the form of the technology we have today.

That form—solid-state and silicon-based—follows one of the most famous axioms in technology: Moore's Law. It relates that the number of transistors that can be fabricated on a silicon integrated circuit—and therefore the computing speed of such a circuit—is doubling every 18 to 24 months. After following this remarkable curve for four decades, solid-state microelectronics has advanced to the point at which engineers can now put on a sliver of silicon of just a few square centimeters some 100 million transistors, with key features measuring 0.18 micron.

These transistors are still far larger than molecular-scale devices. To put the size differential in perspective, if the conventional transistor were scaled up so that it occupied the printed page you are reading, a molecular device would be the period at the end of this sentence. Even in a dozen years, when industry projections suggest that silicon transistors will have shrunk to about 120 nanometers in length, they will still be more than 60,000 times larger in area than molecular electronic devices.

Moreover, no one expects conventional silicon-based microelectronics to continue following Moore's Law forever. At some point, chip-fabrication specialists will find it economically infeasible to continue scaling down microelectronics. As they pack more transistors onto a chip, phenomena such as stray signals on the chip, the need to dissipate the heat from so many closely packed devices, and the difficulty of creating the devices in the first place will halt or severely slow progress.

Indeed, various nagging (though not yet fundamental) problems in the fabrication of efficient smaller silicon transistors and their interconnections are becoming increasingly bothersome. Many experts expect these challenges to intensify dramatically as the transistors approach the 0.1-micron level. Because of these and other difficulties, the exponential increase in transistor densities and processing rates of integrated circuits is being sustained only by a similar exponential rise in the financial outlays necessary to build the facilities that produce these chips. Eventually the drive to downscale will run headlong into these extreme facility costs, and the market will reach equilibrium. Many experts project that this will happen around or before 2015, when a fabrication facility is projected to cost nearly $200 billion. When that happens, the long period of breathtaking advances in the processing power of computer chips will have run its course. Further increases in the power of the chips will be prohibitively costly.

Unfortunately, this impasse will almost certainly occur long before computer chips have reached the power to fulfill some of the most sought-after goals in computer science, such as the creation of extremely sophisticated electronic "brains" that will enable robots to perform on a par with humans in intellectual and cognitive tasks.

## Billions and Billions

The extraordinarily small size of molecular devices brings advantages beyond the simple ability to pack more of them into a small area. To grasp these important benefits requires an understanding of how the devices work—which in turn demands some knowledge of how electrons behave when confined to regions as small as atoms and molecules.

Free electrons can take on energy levels from a continuous range of possibilities. But in atoms or molecules, electrons have energy levels that are quantized: they can only be any one of a number of discrete values, like rungs on a ladder. This series of discrete energy values is a consequence of quantum theory and is true for any system in which the electrons are confined to an infinitesimal space. In molecules, electrons arrange themselves as bonds among atoms that resemble dispersed "clouds," called orbitals. The shape of the orbital is determined by the type and geometry of the constituent atoms. Each orbital is a single, discrete energy level for the electrons.

Even the smallest conventional microtransistors in an integrated circuit are still far too large to quantize the electrons within them. In these devices the movement of electrons is governed by physical characteristics—known as band structures—of their constituent silicon atoms. What that means is that the electrons are moving in the material within a band of allowable energy levels that is quite large relative to the energy levels permitted in a single atom or molecule. This large range of allowable energy levels permits electrons to gain enough energy to leak from one device to the next. And when these conventional devices approach the scale of a few hundred nanometers, it becomes extremely difficult to prevent the minute electric currents that represent information from leaking from one device to an adjacent one. In effect, the transistors leak the electrons that represent information, making it difficult for them to stay in the "off" state.

## Building from the Bottom Up

Besides enabling molecular devices to contain their electrons more securely, quantum mechanical phenomena can also be exploited in specially designed molecules to perform other functions. For example, to construct a "wire" we need an elongated molecule through which electrons can flow easily from one end to the other. Electrons in any quantized structure such as a molecule tend to move from higher- to lower-energy levels, so in order to channel electrons we need a molecule that has an empty, low-energy orbital that is dispersed throughout the molecule from one end to the other. A typical empty, low-energy electron orbital is known as a pi orbital. And the configuration in which electron clouds overlap from one molecular component to the next is called conjugated, so our molecular wire is known as a "pi-conjugated system."

An active device such as a transistor, however, has to do more than merely allow electrons to flow—it has to somehow control that flow. Thus, the task of the molecular device engineer is to exploit the quantum world's discrete energy levels—specifically, by designing molecules whose orbital characteristics achieve the desired kind of electronic control. For example, with the right overlap of orbitals in the molecule, electrons flow. But when the overlap is disturbed—because the molecule has been twisted or its geometry has been otherwise affected—the flow is blocked. In other words, the key to control on the molecular scale is manipulating the number of electrons that are allowed to flow at low orbital energy by perturbing the orbital overlap through the molecule.

Already the standard methods of chemical synthesis allow researchers to design and produce molecules with specific atoms, geometries and orbital arrangements. Moreover, enormous quantities of these molecules are created at the same time, all of them absolutely identical and flawless. Such uniformity is extremely difficult and expensive to achieve in other batch-fabrication processes, such as the lithography-based process used to produce the millions of transistors on an integrated circuit.

The methods used to produce molecular devices are the same as those of the pharmaceutical industry. Chemists start with a compound and then gradually transform it by adding prescribed reagents whose molecules are known to bond to others at specific sites. The procedure may take many steps, but gradually the pieces come together to form a new potential molecular device with a desired orbital structure. After the molecules are made, we use analytical technologies such as infrared spectroscopy, nuclear magnetic resonance and mass spectrometry to determine or confirm the structure of the molecules. The various technologies contribute different pieces of information about the molecule, including its molecular weight and the connection point or angle of a certain fragment. By combining the information, we determine the structure after each step as the new molecule is synthesized.

One of our simplest active devices was a molecule based on a string of three benzene rings, in which the orbitals overlapped (were conjugated) throughout. We made the connections between the benzene rings structurally weak, so that slight twists or kinks weakened or strengthened the conjugation of the orbitals. All we needed was a way to control this twisting and we would have a molecular device in which we could control current flow—a switch, in other words.

To the center benzene ring in the molecule, we added $NO_2$ and $NH_2$ groups, projecting outward from the string on opposite sides of the center ring. This asymmetrical configuration left the molecule with a strongly perturbed electron cloud. That asymmetric, perturbed cloud in turn made the molecule very susceptible to distortion by an electric field: applying an electric field to the molecule twisted it. We now had an active device: every time we applied a voltage to the molecule, an electric field was set up that twisted the molecule and blocked current flow. With the voltage removed, the molecule sprang back to its original shape, and the current flowed again. In follow-up experiments, we found that for our infinitesimal device the abruptness of the switching from one state to the other was superior to that of any comparable solid-state device.

Of course, a lot of advanced technology and years of research were necessary before we could even test one of these devices. The basic challenge is reaching into an unfathomably Lilliputian domain in order to contact and interact with a single molecule and bring information about the behavior of that molecule into our macroscopic world.

The task was all but impossible before the invention, in the 1980s, of the scanning tunneling microscope (STM) at IBM's research laboratories in Zurich. The STM gives scientists a window on the atomic world, letting them visualize and manipulate single atoms or molecules. With an atomically sharp tip of metal held precisely over a surface, the topography of the surface is sensed by the minute current of tunneling electrons that flows between the surface and the tip. Rastering the tip back and forth creates a picture of the hills and valleys on the surface.

Although scanning tunneling microscopy is crucial for testing and constructing individual devices, any useful molecular circuit will consist of vast numbers of devices, orderly arranged and securely affixed to a solid structure to keep them from interacting randomly with one another. Progress toward solving this huge challenge has emerged from studies of self-assembly, a phenomenon in which atoms, molecules or groups of molecules arrange themselves spontaneously into regular patterns and even relatively complex systems without intervention from outside.

## Molecular Glue

Once the assembly process has been set in motion, it proceeds on its own to some desired end. In our research we use self-assembly to attach extremely large numbers of molecules to a surface, typically a metal one. When attached, the molecules, which are often elongated in shape, protrude up from the surface, like a vast forest with identical trees spaced out in a perfect array.

Researchers have studied a variety of self-assembly systems. Our work often requires us to attach molecular devices to a metal (usually gold) surface. So we frequently work with a molecular fragment that we attach to one or both ends of our device and that has a high affinity for gold atoms. The specific fragment we commonly use, called a "sticky" end group for obvious reasons, is based on an atom of sulfur and is known in chemical terminology as thiol.

To initiate the self-assembly, we need only dip a gold surface into a beaker. In solution in this container are our molecular devices, each with thiol end groups on both ends. Spontaneously and in unimaginably large numbers, the devices attach themselves to the gold surface.

Handy though it is, self-assembly alone will not suffice to produce useful molecular-computing systems, at least not initially. For some time, we will have to combine self-assembly with fabrication methods, such as photolithography, borrowed from conventional semiconductor manufacturing. In photolithography, light or some other form of electromagnetic radiation is projected through a stencil-like mask to create patterns of metal and semiconductor on the surface of a semiconducting wafer. In our research we use photolithography to generate layers of metal interconnections and also holes in deposited insulating material. In the holes, we create the electrical contacts and selected spots where molecules are constrained to self-assemble. Thus, the final system consists of regions of self-assembled molecules attached by a mazelike network of metal interconnections.

The first successful demonstration of self-assembly in molecular electronics occurred in 1996, when Paul S. Weiss's group at Pennsylvania State University tested self-assembled molecules. One of us (Tour), then at the University of South Carolina, synthesized the devices. Weiss and his colleagues found that by mixing a small amount of a solution of molecules that were designed to have conducting properties with another containing a known inert insulating molecule, they could get a self-assembled layer in which conductive molecules were very sparsely interspersed among nonconductive ones. By positioning the tip of an STM directly over one of the isolated conducting molecules, they could qualitatively measure the conductivity. As expected, it was significantly greater than that of the surrounding molecules. Similar results were also obtained by a group at Purdue University, which tagged the top of the conductive molecules with minute gold particles.

At the same time at Yale, one of us (Reed) performed the first quantitative electrical measurements of a single molecule, which was also fabricated by self-assembly. Specifically, Reed and his group measured how much current could flow across a single molecule. The heart of the experimental setup was an STM modified to enable it to position two tips opposite each other with sufficient precision and mechanical stability to contain a single molecule in between. A very simple molecule was

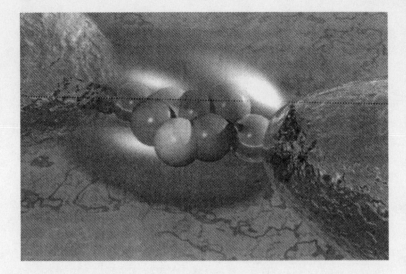

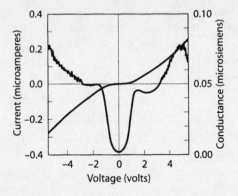

The key to successful molecular circuitry is the ability to control the flow of current. Tests suggest that a benzendithiol molecule – basically a benzene ring modified to interface with a conductive surface, pictured at top — could play the part as the molecule can channel through itself roughly a million million ($10^{12}$) electrons per second. Also encouraging is the molecule's ability to handle the relatively large current flow (graph).

used to convey mobile electrons: a single benzene ring with sticky thiol end groups on both ends to contact the metal leads of the STM tips. It turned out that the resistance of the molecule was in the range of tens of millions of ohms.

The Yale researchers also found that the molecule could sustain a current of about 0.2 microampere at five volts—which meant that the molecule could channel through itself roughly a million million ($10^{12}$) electrons per second. The number is impressive—all the more so in light of the fact that the electrons can pass through the molecule only in single file (one at a time). The magnitude of the current was far larger than would be expected from simple calculations of the power dissipated in a molecule, leading to the conclusion that the electrons traveled through the molecule without generating heat by interacting or colliding.

These initial observations of conduction in molecules were followed quickly by demonstrations of basic devices. The simplest electronic device is a diode, which can be thought of as a one-way valve for electrons. In 1997, only a year after the first measurements of conduction in molecules, two separate research groups built diodes. At the University of Alabama, Robert M. Metzger's group synthesized a molecule that had an internal energetic lineup of orbitals, which varied depending on the polarity of the voltage applied to it. The lineup of orbitals was analogous to the rungs on a ladder. With the voltage applied in one direction, the lineup corresponded to a ladder propped against a house. In this orientation, it takes considerable effort to climb the ladder. With the opposite voltage polarity, the orbital lineup was analogous to the rungs of a ladder lying flat on the ground, where it can be traversed with little effort.

In the other group at Yale, Chong-Wu Zhou took a slightly different tack. With this molecular diode, the differences in the lineup of the energy levels occurred externally to the molecule, where it contacted the metal. This scheme also worked well and helped to set the stage for the design of more useful and interesting molecular devices and circuits.

## Connecting from the Top Down

As they began constructing such devices, the Yale group adapted a structure first made by Kristin Ralls and Robert A. Buhrman of Cornell University. The structure contained an extremely minute hole, called a nanopore, in which an "active region" was created by self-assembling a relatively small number of molecular devices in a single layer, or monolayer. In a hole just 30 nanometers wide, approximately 1,000 of the molecular devices were allowed to self-assemble. Evaporating a metal contact onto the top of the self-assembled monolayer ("SAM") completed the device.

After using this configuration to produce and test molecular diodes, the Yale group quickly moved on to more complex devices, namely, switches. A controllable switch of some kind is a minimum requirement for a general-purpose computer. Even more desirable is a switch that can amplify a current, besides merely turning it on and off. Such amplification is necessary to connect vast numbers of the switches, as is required to build complex logic circuits. The silicon transistor fulfills both those requirements, which is why it is one of the great success stories of the 20th century.

The molecular equivalent of a transistor that can both switch and amplify current is yet to be discovered. But researchers have taken the first steps along the path by constructing switches, such as the twisting switch described earlier. In fact, Jia Chen, a graduate student in Reed's Yale group, observed impressive switching characteristics, such as an on/off ratio greater than 1,000, as measured by the current flow in the two different states. For comparison, the analogous device in the solid-state world, called a resonant tunneling diode, has an on/off ratio of around 100.

Similar behavior was observed in the U.C.L.A./HP experiments. In their demonstration, they showed that the conductivity of a molecular layer of rotaxanes, molecules that resemble a core with a surrounding barbell, could be predictably interrupted when a high voltage was applied to a junction containing the molecules. At this voltage, the molecules reacted and changed configuration, altering the lineup of orbitals and interrupting the flow of current through the molecule. Combining a series of these junctions, they built a device that performed a simple logic function.

Perhaps most encouragingly, molecular devices have already proved themselves as memory elements. Besides active, transistorlike devices, memory is the other main requirement for a useful, general-purpose computer. The construction of the memory element, which involved a relatively straightforward modification to the twisting switch, also demonstrated the ease and flexibility in which molecular-scale devices can be redesigned.

Given the enormous potential advantages of molecular devices, why don't we scrap silicon research and proceed wholeheartedly to molecular-based systems? Because despite the recent auspicious advances, a number of significant obstacles, some fundamental, still stand in the way of fabulously complex and powerful circuits.

## Needed: The Next Transistor

Foremost among them is the challenge of making a molecular device that operates analogously to a transistor. A transistor has three terminals, one of which controls the current flow between the other two. Effective though it was, our twisting switch had only two terminals, with the current flow controlled by an electrical field. In a field-effect transistor, the type in an integrated circuit, the current is also controlled by an electrical field. But the field is set up when a voltage is applied to the third terminal.

A three-terminal molecular device will make possible the chemical synthesis of tremendously efficient and complex circuits. Even before then, combinations of molecular systems with conventional electronics will probably be used in places where the advantages of self-assembly are natural. But interfacing between the molecular and microelectronic worlds will present its own challenges. Computer chips today have two levels of size scale. From the macroscopic level of the chip we can see and hold in our hand, there is a factor of 1,000 in size reduction to get to the gross wiring level, encompassing the largest connections on the chip, which are smaller than a human hair. Then another factor-of-1,000 reduction is necessary to get to the level of the smallest connections and components of the transistors. If molecular devices are to be added to a chip, they will represent yet another factor-of-1,000 reduction in scale down from the smallest microelectronic device components.

Thermal challenges are also staggering, especially if engineers wind up with no alternatives to using molecular devices in modes and configurations similar to those used now with transistors in conventional chips. At present, a state-of-the-art microprocessor with 10 million transistors and a clock cycle of half a gigahertz (half a billion cycles per second) emits almost 100 watts—greater in radiant heat than a range-top cooking surface in the home. Such a unit is close to the thermal limitation of semiconductor technology. Knowing the minimum amount of heat that a single molecular device emits would help put a limit on the number of devices we could put on a chip or substrate of some kind.

This fundamental limit of a molecule, operating at room temperature and at today's speeds, is about 50 picowatts (50 millionths of a millionth of a watt). That figure suggests an upper limit to the number of molecular devices we can closely aggregate: it is roughly 100,000 times more that what we can now do with silicon microtransistors on a chip. Although that may seem like a vast improvement, it is still far below the density that would be possible if we did not have to worry about heat.

Right now no one knows how to create such an interconnect structure on the molecular level. Straightforward extensions of the present techniques we employ to fabricate complex microelectronics are not practical for molecular-scale electronics, because the lithography needed for creating the interconnections to single molecules is far beyond the capability of known technologies. Is the ability to address every device, the common architecture we use today, necessary or efficient at molecular-scale densities? What will large-scale circuits of this technology look like? Can we use nanotubes, single-walled structures of carbon with diameters of one or two nanometers and lengths of less than a micron, as the next generation of interconnects between molecular-scale devices?

Decades from now, radical departures from present computing design will probably be needed to exploit molecular computing systems fully if we are to extend electronics significantly beyond Moore's Law. We have only very limited ideas about what these departures might be. The ability to construct complex molecular devices, with new paradigms and lists of rules about connecting the various devices, will open up an entirely different way to think about computer design.

Although such departures are fraught with problems, we have no alternative but to solve them if electronics is to continue advancing at something like its current pace well into the next century. And difficult though the challenges may be, the rewards for those who solve the problems could be staggering. By pushing Moore's Law past the limits of the tremendously powerful technology we already have, these researchers will take electronics into vast, uncharted terrain. If we can get to that region, we will almost certainly find some wondrous things—maybe even the circuitry that will give rise to our intellectual successor.

*Intelligent materials—materials that can "sense" their environment and then react to it—seem like the most obvious extension of artificial intelligence research. When the material need only respond within the area of its programmed expertise, success rates and practical applications climb.*

# Intelligent Materials

## by Craig A. Rogers

I magine, for a moment, music in your room or car that emanates from the doors, floor or ceiling; ladders that alert us when they are overburdened and may soon collapse under the strain; buildings and bridges that reinforce themselves during earthquakes and seal cracks of their own accord. Like living beings, these systems would alter their structure, account for damage, effect repairs and retire—gracefully, one hopes—when age takes its toll.

Such structures may seem far-fetched. But, in fact, many researchers have demonstrated the feasibility of such "living" materials. To animate an otherwise inert substance, modern-day alchemists enlist a variety of devices: actuators and motors that behave like muscles; sensors that serve as nerves and memory; and communications and computational networks that represent the brain and spinal column. In some respects, the systems have features that can be considered superior to biological functions—some substances can be hard and strong one moment but made to act like Jell-O the next.

These so-called intelligent materials systems have substantial advantages over traditionally engineered constructs. Henry Petroski, in his book *To Engineer Is Human*, perhaps best articulated the traditional principles. A skilled designer always considers the worst-case scenario. As a result, the design contains large margins of safety, such as numerous reinforcements, redundant subunits, backup subsystems and added mass. This approach, of course, demands more natural resources than are generally required and consumes more energy to produce and maintain a structure. It also expends more human effort trying to predict those circumstances under which an engineered artifact will be used and abused.

Trying to anticipate the worst case has a much more serious and obvious flaw, one we read about in the newspapers and hear about on the evening news from time to time: that of being unable to foresee all possible contingencies. Adding insult to injury is the costly litigation that often ensues.

Intelligent materials systems, in contrast, would avoid most of these problems. Made for a given purpose, they would also be able to modify their behavior under dire circumstances. As an example, a ladder that is overloaded with weight could use electrical energy to stiffen and alert the user of the problem. The overload response would be based on the actual life experience of the ladder, to account for aging or damage. As a result, the ladder would be able to evaluate its current health; when it could no longer perform even minimal tasks, the ladder would announce its retirement. In a way, then, the ladder resembles living bone, which remodels itself under changing loads. But unlike bone, which begins to respond within minutes of an impetus but may take months to complete its growth, an intelligent ladder needs to change in less than a second.

## Muscles for Intelligent Systems

Materials that allow structures such as ladders to adapt to their environment are known as actuators. They can change shape, stiffness, position, natural frequency and other mechanical characteristics in response to temperature or electromagnetic fields. The four most common actuator materials being used today are shape-memory alloys, piezoelectric ceramics, magnetostrictive materials and electrorheological and magnetorheological fluids. Although not one of these categories stands as the perfect artificial muscle, each can nonetheless fulfill particular requirements of many tasks.

Shape-memory alloys are metals that at a certain temperature revert back to their original shape after being strained. In the process of returning to their "remembered" shape, the alloys can generate a large force useful for actuation. Most prominent among them perhaps is the family of the nickel-titanium alloys developed at the Naval Ordnance Laboratory (now the Naval Surface Warfare Center). The material, known as Nitinol (Ni for nickel, Ti for titanium and NOL for Naval Ordnance Lab), exhibits substantial resistance to corrosion and fatigue and recovers well from large deformations. Strains that elongate up to 8 percent of the alloy's length can be reversed by heating the alloy, typically with electric current.

Japanese engineers are using Nitinol in micromanipulators and robotics actuators to mimic the smooth motions of human muscles. The controlled force exerted when the Nitinol recovers its shape allows these devices to grasp delicate paper cups filled with water. Nitinol wires embedded in composite materials have also been used to modify vibrational characteristics. They do so by altering the rigidity or state of stress in the structure, thereby shifting the natural frequency of the composite. Thus, the structure would be unlikely to resonate with any external vibrations, a process known to be powerful enough to bring down a bridge. Experiments have shown that embedded Nitinol can apply compensating compression to reduce stress in a structure. Other applications for these actuators include engine mounts and suspensions that control vibration.

The main drawback of shape-memory alloys is their slow

rate of change. Because actuation depends on heating and cooling, they respond only as fast as the temperature can shift.

A second kind of actuator, one that addresses the sluggishness of the shape-memory alloys, is based on piezoelectrics. This type of material, discovered in 1880 by French physicists Pierre and Jacques Curie, expands and contracts in response to an applied voltage. Piezoelectric devices do not exert nearly so potent a force as shape-memory alloys; the best of them recover only from less than 1 percent strain. But they act much more quickly, in thousandths of a second. Hence, they are indispensable for precise, high-speed actuation. Optical tracking devices, magnetic heads and adaptive optical systems for robots, ink-jet printers and speakers are some examples of systems that rely on piezoelectrics. Lead zirconate titanate (PZT) is the most widely used type.

Recent research has focused on using PZT actuators to attenuate sound, dampen structural vibrations and control stress. At Virginia Polytechnic Institute and State University, piezoelectric actuators were used in bonded joints to resist the tension near locations that have a high concentration of strain. The experiments extended the fatigue life of some components by more than an order of magnitude.

A third family of actuators is derived from magnetostrictive materials. This group is similar to piezoelectrics except that it responds to magnetic, rather than electric, fields. The magnetic domains in the substance rotate until they line up with an external field. In this way, the domains can expand the material. Terfenol-D, which contains the rare earth element terbium, expands by more than 0.1 percent. This relatively new material has been used in low-frequency, high-power sonar transducers, motors and hydraulic actuators. Like Nitinol, Terfenol-D is being investigated for use in the active damping of vibrations.

The fourth kind of actuator for intelligent systems is made of special liquids called electrorheological and magnetorheological fluids. These substances contain micron-size particles that form chains when placed in an electric or magnetic field, resulting in increases in apparent viscosity of up to several orders of magnitude in milliseconds. Applications that have been demonstrated with these fluids include tunable dampers, vibration-isolation systems, joints for robotic arms, and frictional devices such as clutches, brakes and resistance controls on exercise equipment. Still, several problems plague these fluids, such as abrasiveness and chemical instability, and much recent work to improve them is aimed at the magnetic substances.

## Nerves of Glass

Providing the actuators with information are the sensors, which describe the physical state of the materials system. Advances in micromachining have created a wealth of promising electromechanical devices that can serve as sensors. I will focus on two types that are well developed now and are the most likely to be incorporated in intelligent systems: optical fibers and piezoelectric materials.

Optical fibers embedded in a "smart" material can provide data in a couple of ways. First, they can simply provide a steady light signal to a sensor; breaks in the light beam indicate a structural flaw that has snapped the fiber. The second, more subtle, approach involves looking at key characteristics of the light—intensity, phase, polarization or a similar feature. The National Aeronautics and Space Administration and other research centers have used such a fiber-optic system to measure the strain in composite materials. Fiber-optic sensors can also measure magnetic fields, deformations, vibrations and acceleration. Resistance to adverse environments and immunity to electrical or magnetic noise are among the advantages of optical sensors.

In addition to serving as actuators, piezoelectric materials make good sensors. Piezoelectric polymers, such as polyvinylidene fluoride (PVDF), are commonly exploited for sensing because they can be formed in thin films and bonded to many kinds of surfaces. The sensitivity of PVDF films to pressure has proved suitable for sensors tactile enough to read Braille and to distinguish grades of sandpaper. Ultrathin PVDF films, perhaps 200 to 300 microns thick, have been proposed for use in robotics. Such a sensor might replicate the capabilities of human skin, detecting temperature and geometric features such as edges and corners or distinguishing between different fabrics.

Actuators and sensors are crucial elements in an intelligent materials system, but the essence of this new design philosophy rests in the manifestation of the most critical of life functions, intelligence. The exact degree of intelligence—the extent to which the material should be smart or merely adaptive—is debatable. At minimum, there must be an ability to learn about the environment and live within it.

The thinking features that the intelligent materials community is trying to create have constraints that the engineering world has never experienced before. Specifically, the tremendous number of sensors, actuators and their associated power sources would argue against feeding all these devices into a central processor. Instead designers have taken clues from nature. Neurons are not nearly so fast as modern-day silicon chips, but they can nonetheless perform complex tasks with amazing speed because they are networked efficiently.

The key appears to be a hierarchical architecture. Signal processing and the resulting action can take place at levels below and far removed from the brain. The reflex of moving your hand away from a hot stove, for example, is organized entirely within the spinal cord. Less automatic behaviors are organized by successively higher centers within the brain. Besides being efficient, such an organization is fault-tolerant: unless there is some underlying organic reason, we rarely experience a burning sensation when holding an iced drink.

The brains behind an intelligent materials system follow a similar organization. In fact, investigators take their cue from research into artificial life, an out-growth of the cybernetics field. Among the trendiest control concepts is the artificial neural network, which is computer programming that mimics the functions of real neurons. Such software can learn, change in response to contingencies, anticipate needs and correct mistakes—more than adequate functions for intelligent materials systems. Ultimately, computational hardware and the processing algorithms will determine how complex these systems can become—that is, how many sensors and actuators we can use.

## Brains over Brawn

Engineers are incorporating intelligent materials systems into several areas. NASA uses electroactive materials crafted by researchers at Pennsylvania State University to modify the optics of the Hubble Space Telescope.

Perhaps the most mature application at the moment is the control of acoustics. The objective is to reduce sound, be it noise inside an aircraft fuselage shaken by engines or the acoustic signature of a submarine. One way to control noise, of course, is to use brute force. Simply add enough mass to stop the structure from vibrating. In contrast, the intelligent materials approach is to sense the structural oscillations that are radiating the noise and use the actuators distributed throughout the structure to control the most obnoxious vibrations. This concept is the foundation for sound-cancellation headphones used by pilots (and now available through airline gift catalogues), and full systems are being tested on turboprop commuter aircraft.

How far can engineers take intelligent materials? The future lies in developing a system that can behave in more complex ways. For instance, intelligent structures now being demonstrated come with many more sensors than are needed by any one application.

A prospective design might rely on an adaptive architecture in which the sensors can be connected to create the specific system desired. Moreover, the design would be highly flexible. If a particular sensor fails, the adaptive architecture would be able to replace the failed sensor with the next best alternative and reconfigure the interconnections and the control algorithm to accommodate this change.

This level of sophistication would clearly tax the manufacturing process. Large arrays of sensors, actuators, power sources and control processors will most likely require three-dimensional interconnections. Such complexity could easily render a smart structure too expensive to build. One solution to forming complex features cheaply is to rely on techniques of computer-chip makers: photolithography. The process, akin to photocopying, can in principle mass-produce components for fractions of a cent per device. A sensor network might therefore resemble the detail of a silicon microchip.

## A New Way of Engineering

Intelligent systems may not only initiate a materials revolution but may also lead to the next step in our understanding of complex physical phenomena. In many ways, they are the ideal recording devices. They can sense their environments, store detailed information about the state of the material over time and "experiment" on the phenomena by changing properties.

The most lasting influence, however, will be on the philosophy of design. Engineers will not have to add mass and cost to ensure safety. They will learn not from the autopsies performed on structures that have failed but from the actual experiences of the edifice. We will soon have the chance to ask structures how they feel, where they hurt, if they have been abused recently. They might even be able to identify the abuser.

Will smart materials systems eliminate all catastrophic failures? Not any more than trees will stop falling in hurricane winds or birds will no longer tumble when they hit glass windows. But intelligent materials systems will enable inanimate objects to become more natural and lifelike. They will be manifestations of the next engineering revolution—the dawn of a new materials age.

## Here's Looking at You

Parties have a way of generating outrageous ideas. Most don't survive the night, but a scheme that bubbled to the surface at a 1992 event held by Rodney A. Brooks of the Massachusetts Institute of Technology is changing the way researchers think about thinking. Brooks, the head of M.I.T.'s artificial intelligence laboratory, was celebrating the switch-on date of the fictitious Hal 9000 computer, which appeared in the movie 2001: *A Space Odyssey*. As he reflected that no silicon brain could yet rival Hal's slick mendacity, he was seized by the notion of building a humanoid robot based on biological principles, rather than on conventional approaches to robot design.

The robot, known as Cog, started to take shape in the summer of 1993. The project, which was initially to last five years, is intended to reveal problems that emerge in trying to design a humanoid machine and thereby elucidate principles of human cognition. Instead of being programmed with detailed information about its environment and then calculating how to achieve a set goal—the modus operandi of industrial robots—Cog learns about itself and its environment by trial and error. Brooks says that although there are no near-term practical goals for Cog technology, it has stimulated "a bunch" of papers.

Central to the plan was that the robot should (unlike Hal) look and move something like a human being, to encourage people to interact with it. Tufts University philosopher Daniel C. Dennett, an informal adviser to the fluid group of M.I.T. researchers who have worked on Cog, has stated that the machine "will be conscious if we get done all the things we've got written down." Another principle guiding the project was that it should not include a preplanned, or explicit, internal "model" of the world. Rather the changes in Cog as it learns are, in the team's words, "meaningless without interaction with the outside world."

A little after the five-year mark, not even the most enthusiastic fan could argue that Cog is conscious. Yet it is also clear that the exercise of building it has highlighted some intriguing observations.

One day in the fall of '99 Brian Scassellati and Cynthia Breazeal of Brooks's department exhibited some of Cog's tricks. The machine's upper-torso humanoid form is irresistibly reminiscent of C3PO of *Star Wars* fame. It has learned how to turn to fixate on a moving object, first switching its swiveling eyes, then moving its whole head to catch up. Cog will imitate a nod and reach out to touch things with strikingly lifelike arm movements. The movements have a fluidity not usually associated with machines, because they are driven by a system that has learned to exploit the limbs' natural dynamics. Cog's mechanical facility is revealed in the way it quickly picks up the timing needed to play with a slinky toy attached to its hands or spontaneously rotates a crank.

Plans are underway to provide the robot with more tactile sensors, a better controlled posture and the ability to distinguish different sound sources. Cog should then be able to associate a voice with a human being in its visual field. There are no plans to add a premade speech-recognition capability, because that would violate the guiding philosophy that Cog should learn on its own.

An expandable stack of high-speed processors gives Cog enough computing power to build on its current skills, Brooks explains. Yet even in its present, simple incarnation, Cog can elicit unexpected behavior from humans. Breazeal once found herself taking turns with Cog passing an eraser between them, a game she had not planned but which the situation seemed to invite.

Breazeal is now studying emotional interactions with a disembodied Cog-type head equipped with expressive mobile eyelids, ears and a jaw. This robot, called Kismet, might yield insights that will expand Cog's mental horizons. Kismet, unlike Cog, has built-in drives for social activity, stimulation and fatigue and can create expressions of happiness, sadness, anger, fear or disgust. Like a baby, it can manipulate a soft-hearted human into providing it with a companionable level of interaction.

It is clear that Cog is still some years from mastering more sophisticated behaviors. Integrating its subbehaviors so they do not compete is a difficulty that has hardly yet been faced. And Cog has no sense of time. Finding a good way to provide one is a "real challenge," Brooks's team writes. Because the design philosophy requires that Cog function like a human, a digital clock is not acceptable.

Cog's development, it seems, will prove slower than that of a human infant. Perhaps just as well: the team has started to consider the complications that might follow from giving Cog a sense of sexual identity. But the effort to make a machine that acts like a human could yet tell researchers a good deal about how a human acts that way.

—Tim Beardsley in Cambridge, Mass.

*When computing speed matches stride with the speed of the human brain, robotic learning may be as simple as downloading a file. If this happens, the result could be a non-biological entity that displays many of the characteristics of consciousness and intelligence.*

# The Coming Merging of Mind and Machine

## Ray Kurzweil

Sometime early in the next century, the intelligence of machines will exceed that of humans. Within several decades, machines will exhibit the full range of human intellect, emotions and skills, ranging from musical and other creative aptitudes to physical movement. They will claim to have feelings and, unlike today's virtual personalities, will be very convincing when they tell us so. By 2019 a $1,000 computer will at least match the processing power of the human brain. By 2029 the software for intelligence will have been largely mastered, and the average personal computer will be equivalent to 1,000 brains.

Within three decades, neural implants may be available that interface directly to our brain cells. The implants would enhance sensory experiences and improve our memory and thinking. Once computers achieve a level of intelligence comparable to that of humans, they will necessarily soar past it. For example, if I learn French, I can't readily download that learning to you. The reason is that for us, *learning* involves successions of stunningly complex patterns of interconnections among brain cells (neurons) and among the concentrations of biochemicals, known as neurotransmitters, that enable impulses to travel from neuron to neuron. We have no way of quickly downloading these patterns. But quick downloading will allow our nonbiological creations to share immediately what they learn with billions of other machines. Ultimately, nonbiological entities will master not only the sum total of their own knowledge but all of ours as well.

As this happens, there will no longer be a clear distinction between human and machine. We are already putting computers—*neural implants*—directly into people's brains to counteract Parkinson's disease and tremors from multiple sclerosis. We have *cochlear implants* that restore hearing. A *retinal implant* is being developed in the U.S. that is intended to provide at least some visual perception for some blind individuals, basically by replacing certain visual-processing circuits of the brain. Recently scientists from Emory University implanted a chip in the brain of a paralyzed stroke victim that allows him to use his brainpower to move a cursor across a computer screen.

In the 2020s neural implants will improve our sensory experiences, memory and thinking. By 2030, instead of just phoning a friend, you will be able to meet in, say, a virtual Mozambican game preserve that will seem compellingly real. You will be able to have any type of experience—business, social, sexual—with anyone, real or simulated, regardless of physical proximity.

## How Life and Technology Evolve

To gain insight into the kinds of forecasts I have just made, it is important to recognize that technology is advancing exponentially. An exponential process starts slowly, but eventually its pace increases extremely rapidly. (A fuller documentation of my argument is contained in my book, *The Age of Spiritual Machines*.)

The evolution of biological life and the evolution of technology have both followed the same pattern: they take a long time to get going, but advances build on one another and progress erupts at an increasingly furious pace. We are entering that explosive part of the technological evolution curve right now.

Consider: It took billions of years for *Earth to form*. It took two billion more for *life to begin* and almost as long for molecules to organize into the *first multicellular plants* and *animals* about 700 million years ago. The pace of evolution quickened as *mammals* inherited Earth some 65 million years ago. With the emergence of primates, evolutionary progress was measured in mere millions of years, leading to *Homo sapiens* perhaps 500,000 years ago.

The evolution of technology has been a continuation of the evolutionary process that gave rise to us—the technology-creating species—in the first place. It took tens of thousands of years for our ancestors to figure out that sharpening both sides of a stone created useful tools. Then, earlier in this millennium, the time required for a major paradigm shift in technology had shrunk to hundreds of years.

The pace continued to accelerate during the 19th century, during which technological progress was equal to that of the 10 centuries that came before it. Advancement in the first two decades of the 20th century matched that of the entire 19th century. Today significant technological transformations take just a few years; for example, the World Wide Web, already a ubiquitous form of communication and commerce, did not exist prior to 1990.

Computing technology is experiencing the same exponential growth. Over the past several decades, a key factor in this expansion has been described by Moore's Law. Gordon Moore, a co-founder of Intel, noted in the mid-1960s that technologists had been doubling the density of transistors on integrated circuits every 12 months. This meant computers were periodically doubling both in capacity and in speed per unit cost. In the mid-1970s Moore revised his observation of the doubling time to a more accurate estimate of about 24 months, and that trend has persisted through the 1990s.

After decades of devoted service, Moore's Law will have run its course around 2019. By that time, transistor features will be just a few atoms in width. But new computer architectures will continue the exponential growth of computing. For example, computing cubes are already being designed that will provide thousands of layers of circuits, not just one as in today's computer chips. Other technologies that promise orders-of-magnitude increases in computing density include *nanotube circuits* built from carbon atoms, *optical computing,* crystalline computing and *molecular computing.*

We can readily see the march of computing by plotting the speed (in instructions per second) per $1,000 (in constant dollars) of 49 famous calculating machines spanning the 20th century. The graph is a study in exponential growth: computer speed per unit cost doubled every three years between 1910 and 1950 and every two years between 1950 and 1966 and is now doubling every year. It took 90 years to achieve the first $1,000 computer capable of executing one million instructions per second (MIPS). Now we add an additional MIPS to a $1,000 computer every day.

## Why Returns Accelerate

Why do we see exponential progress occurring in biological life, technology and computing? It is the result of a fundamental attribute of any evolutionary process, a phenomenon I call the Law of Accelerating Returns. As order exponentially increases (which reflects the essence of evolution), the time between salient events grows shorter. Advancement speeds up. The returns—the valuable products of the process—accelerate at a nonlinear rate. The escalating growth in the price performance of computing is one important example of such accelerating returns.

A frequent criticism of predictions is that they rely on an unjustified extrapolation of current trends, without considering the forces that may alter those trends. But an evolutionary process accelerates because it builds on past achievements, including improvements in its own means for further evolution. The resources it needs to continue exponential growth are its own increasing order and the chaos in the environment in which the evolutionary process takes place, which provides the options for further diversity. These two resources are essentially without limit.

The Law of Accelerating Returns shows that by 2019 a $1,000 personal computer will have the processing power of the human brain—20 million billion calculations per second. Neuroscientists came up with this figure by taking an estimation of the number of neurons in the brain, 100 billion, and multiplying it by 1,000 connections per neuron and 200 calculations per second per connection. By 2055, $1,000 worth of computing will equal the processing power of all human brains on Earth (of course, I may be off by a year or two). Computer power per unit cost is now doubling every year.

## Programming Intelligence

That's the prediction for processing power, which is a necessary but not sufficient condition for achieving human-level intelligence in machines. Of greater importance is the software of intelligence.

One approach to creating this software is to painstakingly program the rules of complex processes. We are getting good at this task in certain cases; the CYC (as in "encyclopedia") system designed by Douglas B. Lenat of Cycorp has more than one million rules that describe the intricacies of human common sense, and it is being applied to Internet search engines so that they return smarter answers to our queries.

Another approach is "complexity theory" (also known as chaos theory) computing, in which self-organizing algorithms gradually learn patterns of information in a manner analogous to human learning. One such method, *neural nets*, is based on simplified mathematical models of mammalian neurons. Another method, called genetic (or evolutionary) algorithms, is based on allowing intelligent solutions to develop gradually in a simulated process of evolution.

Ultimately, however, we will learn to program intelligence by copying the best intelligent entity we can get our hands on: the human brain itself. We will reverse-engineer the human brain, and fortunately for us it's not even copyrighted!

The most immediate way to reach this goal is by destructive scanning: take a brain frozen just before it was about to expire and examine one very thin slice at a time to reveal every neuron, interneuronal connection and concentration of neurotransmitters across each gap between neurons (these gaps are called synapses). One condemned killer has already allowed *his brain and body* to be scanned, and all 15 billion bytes of him can be accessed on the National Library of Medicine's Web site. The resolution of these scans is not nearly high enough for our purposes, but the data at least enable us to start thinking about these issues.

We also have noninvasive scanning techniques, including high-resolution magnetic resonance imaging (MRI) and others. Their increasing resolution and speed will eventually enable us to resolve the connections between neurons. The rapid improvement is again a result of the Law of Accelerating Returns, because massive computation is the main element in higher-resolution imaging.

Another approach would be to send microscopic robots (or "nanobots") into the bloodstream and program them to explore every capillary, monitoring the brain's connections and neurotransmitter concentrations.

## Fantastic Voyage

Although sophisticated robots that small are still several decades away at least, their utility for probing the innermost recesses of our bodies would be far-reaching. They would communicate wirelessly with one another and report their findings to other computers. The result would be a noninvasive scan of the brain taken from within.

Most of the technologies required for this scenario already exist, though not in the microscopic size required. Miniaturizing them to the tiny sizes needed, however, would reflect the essence of the Law of Accelerating Returns. For example, the translators on an integrated circuit have been shrinking by a factor of approximately 5.6 in each linear dimension every 10 years.

The capabilities of these embedded nanobots would not be limited to passive roles such as monitoring. Eventually they could be built to communicate directly with the neuronal circuits in our brains, enhancing or extending our mental capabilities. We already have electronic devices that can communicate with neurons by detecting their activity and either triggering nearby neurons to fire or suppressing them from firing. The embedded nanobots will be capable of reprogramming neural connections to provide virtual-reality experiences and to enhance our pattern recognition and other cognitive faculties.

To decode and understand the brain's information-processing methods (which, incidentally, combine both digital and analog methods), it is not necessary to see every connection, because there is a great deal of redundancy within each region. We are already applying insights from early stages of this reverse-engineering process. For example, in speech recognition, we have already decoded and copied the brain's early stages of sound processing.

Perhaps more interesting than this scanning-the-brain-to-understand-it approach would be scanning the brain for the purpose of downloading it. We would map the locations, interconnections, and contents of all the neurons, synapses and neurotransmitter concentrations. The entire organization, including the brain's memory, would then be re-created on a digital-analog computer.

To do this, we would need to understand local brain processes, and progress is already under way. Theodore W. Berger and his co-workers at the University of Southern California have built integrated circuits that precisely match the processing characteristics of substantial clusters of neurons. Carver A. Mead and his colleagues at the California Institute of Technology have built a variety of integrated circuits that emulate the digital-analog characteristics of mammalian neural circuits.

Developing complete maps of the human brain is not as daunting as it may sound. The Human Genome Project seemed impractical when it was first proposed. At the rate at which it was possible to scan genetic codes in the 1980s, it would have taken thousands of years to complete the genome. But in accordance with the Law of Accelerating Returns, the ability to sequence DNA has been accelerating. The latest estimates are that the entire human genome will be completed in just a few years.

By the third decade of the 21st century, we will be in a position to create complete, detailed maps of the computationally relevant features of the human brain and to re-create these designs in advanced neural computers. We will provide a variety of bodies for our machines, too, from virtual bodies in virtual reality to bodies comprising swarms of nanobots. In fact, *humanoid robots* that ambulate and have lifelike facial expressions are already being developed at several laboratories in Tokyo.

## Will It Be Conscious?

Such possibilities prompt a host of intriguing issues and questions. Suppose we scan someone's brain and reinstate the resulting "mind file" into a suitable computing medium. Will the entity that emerges from such an operation be conscious? This being would appear to others to have very much the same personality, history and memory. For some, that is enough to define consciousness. For others, such as physicist and author James Trefil, no logical reconstruction can attain human consciousness, although Trefil concedes that computers may become conscious in some new way.

At what point do we consider an entity to be *conscious,* to be self-aware, to have free will? How do we distinguish a process that is conscious from one that just acts as *if* it is conscious? If the entity is very convincing when it says, "I'm lonely, please keep me company," does that settle the issue?

If you ask the "person" in the machine, it will strenuously claim to be the original person. If we scan, let's say, me, and reinstate that information into a neural computer, the person who emerges will think he is (and has been) me (or at least he will act that way). He will say, "I grew up in Queens, New York, went to college at M.I.T., stayed in the Boston area, walked into a scanner there and woke up in the machine here. Hey, this technology really works." But wait, is this really me? For one thing, old Ray (that's me) still exists in my carbon-cell-based brain.

Will the new entity be capable of spiritual experiences? Because its brain processes are effectively identical, its behavior will be comparable to that of the person it is based on. So it will certainly claim to have the full range of emotional and spiritual experiences that a person claims to have.

No objective test can absolutely determine consciousness. We cannot objectively measure subjective experience (this has to do with the very nature of the concepts "objective" and "subjective"). We can measure only correlates of it, such as behavior. The new entities will appear to be conscious, and whether or not they actually are will not affect their behavior. Just as we debate today the consciousness of nonhuman entities such as animals, we will surely debate the potential consciousness of nonbiological intelligent entities. From a practical perspective, we will accept their claims. They'll get mad if we don't.

Before the next century is over, the Law of Accelerating Returns tells us, Earth's technology-creating species—*us*—will merge with our own technology. And when that happens, we might ask: What is the difference between a human brain enhanced a millionfold by neural implants and a nonbiological intelligence based on the reverse-engineering of the human brain that is subsequently enhanced and expanded?

The engine of evolution used its innovation from one period (humans) to create the next (intelligent machines). The subsequent milestone will be for the machines to create their own next generation without human intervention.

An evolutionary process accelerates because it builds on its own means for further evolution. Humans have beaten evolution. We are creating intelligent entities in considerably less time than it took the evolutionary process that created us. Human intelligence—a product of evolution—has transcended it. So, too, the intelligence that we are now creating in computers will soon exceed the intelligence of its creators.

*Moving about the world with competency and ease is how most people envision robots. However, it is not the lack of mobility that is thwarting the rise of artificial intelligence—it is the lack of sufficient raw power. Computing power on the order of 500 million million instructions per second needs to be achieved before a robot can approach even an apelike level of intelligence.*

# Rise of the Robots

## Hans Moravec

I n recent years the mushrooming power, functionality and ubiquity of computers and the Internet have outstripped early forecasts about technology's rate of advancement and usefulness in everyday life. Alert pundits now foresee a world saturated with powerful computer chips, which will increasingly insinuate themselves into our gadgets, dwellings, apparel and even our bodies.

Yet a closely related goal has remained stubbornly elusive. In stark contrast to the largely unanticipated explosion of computers into the mainstream, the entire endeavor of robotics has failed rather completely to live up to the predictions of the 1950s. In those days, experts who were dazzled by the seemingly miraculous calculational ability of computers thought that if only the right software were written, computers could become the artificial brains of sophisticated autonomous robots. Within a decade or two, they believed, such robots would be cleaning our floors, mowing our lawns and, in general, eliminating drudgery from our lives.

Obviously, it hasn't turned out that way. It is true that industrial robots have transformed the manufacture of automobiles, among other products. But that kind of automation is a far cry from the versatile, mobile, autonomous creations that so many scientists and engineers have hoped for. In pursuit of such robots, waves of researchers have grown disheartened and scores of start-up companies have gone out of business.

It is not the mechanical "body" that is unattainable; articulated arms and other moving mechanisms adequate for manual work already exist, as the industrial robots attest. Rather it is the computer-based artificial brain that is still well below the level of sophistication needed to build a humanlike robot.

Nevertheless, I am convinced that the decades-old dream of a useful, general-purpose autonomous robot will be realized in the not too distant future. By 2010 we will see mobile robots as big as people but with cognitive abilities similar in many respects to those of a lizard. The machines will be capable of carrying out simple chores, such as vacuuming, dusting, delivering packages and taking out the garbage. By 2040, I believe, we will finally achieve the original goal of robotics and a thematic mainstay of science fiction: a freely moving machine with the intellectual capabilities of a human being.

## Reasons for Optimism

In light of what I have just described as a history of largely unfulfilled goals in robotics, why do I believe that rapid progress and stunning accomplishments are in the offing? My confidence is based on recent developments in electronics and software, as well as on my own observations of robots, computers and even insects, reptiles and other living things over the past 30 years.

The single best reason for optimism is the soaring performance in recent years of mass-produced computers. Through the 1970s and 1980s, the computers readily available to robotics researchers were capable of executing about one million instructions per second (MIPS). Each of these instructions represented a very basic task, like adding two 10-digit numbers or storing the result in a specified location in memory.

In the 1990s computer power suitable for controlling a research robot shot through 10 MIPS, 100 MIPS and has lately reached 1,000 in high-end desktop machines. Thus, functions far beyond the capabilities of robots in the 1970s and 1980s are now coming close to commercial viability.

For example, in October 1995 an experimental vehicle called Navlab V crossed the U.S. from Washington, D.C., to San Diego, driving itself more than 95 percent of the time. The vehicle's self-driving and navigational system was built around a 25-MIPS laptop based on a microprocessor by Sun Microsystems. The Navlab V was built by the Robotics Institute at Carnegie Mellon University, of which I am a member. Similar robotic vehicles, built by researchers elsewhere in the U.S. and in Germany, have logged thousands of highway kilometers under all kinds of weather and driving conditions.

In other experiments within the past few years, mobile robots mapped and navigated unfamiliar office suites, and computer vision systems located textured objects and tracked and analyzed faces in real time. Meanwhile personal computers became much more adept at recognizing text and speech.

Still, computers are no match today for humans in such functions as recognition and navigation. This puzzled experts for many years, because computers are far superior to us in calculation. The explanation of this apparent paradox follows from the fact that the human brain, in its entirety, is not a true programmable, general-purpose computer (what computer scientists refer to as a universal machine; almost all computers nowadays are examples of such machines).

To understand why this is requires an evolutionary perspective. To survive, our early ancestors had to do several things repeatedly and very well: locate food, escape predators, mate and protect offspring. Those tasks depended strongly on the brain's ability to recognize and navigate. Honed by hundreds of millions of years of evolution, the brain became a kind of ultra-sophisticated—but special-purpose—computer.

The ability to do mathematical calculations, of course, was irrelevant for survival. Nevertheless, as language transformed human culture, at least a small part of our brains evolved into a universal machine of sorts. One of the hallmarks of such a machine is its ability to follow an arbitrary set of instructions, and with language, such instructions could be transmitted and carried out. But because we visualize numbers as complex shapes, write them down and perform other such functions, we process digits in a monumentally awkward and inefficient way. We use hundreds of billions of neurons to do in minutes what hundreds of them, specially "rewired" and arranged for calculation, could do in milliseconds.

A tiny minority of people are born with the ability to do seemingly amazing mental calculations. In absolute terms, it's not so amazing: they calculate at a rate perhaps 100 times that of the average person. Computers, by comparison, are millions or billions of times faster.

## Can Hardware Simulate Wetware?

The challenge facing roboticists is to take general-purpose computers and program them to match the largely special-purpose human brain, with its ultraoptimized perceptual inheritance and other peculiar evolutionary traits. Today's robot-controlling computers are much too feeble to be applied successfully in that role, but it is only a matter of time before they are up to the task.

Implicit in my assertion that computers will eventually be capable of the same kind of perception, cognition and thought as humans is the idea that a sufficiently advanced and sophisticated artificial system—for example, an electronic one—can be made and programmed to do the same thing as the human nervous system, including the brain. This issue is controversial in some circles right now, and there is room for brilliant people to disagree.

At the crux of the matter is the question of whether biological structure and behavior arise entirely from physical law and whether, moreover, physical law is computable—that is to say, amenable to computer simulation. My view is that there is no good scientific evidence to negate either of these propositions. On the contrary, there are compelling indications that both are true.

Molecular biology and neuroscience are steadily uncovering the physical mechanisms underlying life and mind but so far have addressed mainly the simpler mechanisms. Evidence that simple functions can be composed to produce the higher capabilities of nervous systems comes from programs that read, recognize speech, guide robot arms to assemble tight components by feel, classify chemicals by artificial smell and taste, reason about abstract matters and so on. Of course, computers and robots today fall far short of broad human or even animal competence. But that situation is understandable in light of an analysis, summarized in the next section, that concludes that today's computers are only powerful enough to function like insect nervous systems. And, in my experience, robots do indeed perform like insects on simple tasks.

Ants, for instance, can follow scent trails but become disoriented when the trail is interrupted. Moths follow pheromone trails and also use the moon for guidance. Similarly, many commercial robots today follow guide wires installed beneath the surface they move over, and some orient themselves using lasers that read bar codes on walls.

If my assumption that greater computer power will eventually lead to human-level mental capabilities is true, we can expect robots to match and surpass the capacity of various animals and then finally humans as computer-processing rates rise sufficiently high. If on the other hand the assumption is wrong, we will someday find specific animal or human skills that elude implementation in robots even after they have enough computer power to match the whole brain. That would set the stage for a fascinating scientific challenge—to somehow isolate and identify the fundamental ability that brains have and that computers lack. But there is no evidence yet for such a missing principle.

The second proposition, that physical law is amenable to computer simulation, is increasingly beyond dispute. Scientists and engineers have already produced countless useful simulations, at various levels of abstraction and approximation, of everything from automobile crashes to the "color" forces that hold quarks and gluons together to make up protons and neutrons.

## Nervous Tissue and Computation

If we accept that computers will eventually become powerful enough to simulate the mind, the question that naturally arises is: What processing rate will be necessary to yield performance on a par with the human brain? To explore this issue, I have considered the capabilities of the vertebrate retina, which is understood well enough to serve as a Rosetta stone roughly relating nervous tissue to computation. By comparing how fast the neural circuits in the retina perform image-processing operations with how many instructions per second it takes a computer to accomplish similar work, I believe it is possible to at least coarsely estimate the information-processing power of nervous tissue—and by extrapolation, that of the entire human nervous system.

The human retina is a patch of nervous tissue in the back of the eyeball half a millimeter thick and approximately two centimeters across. It consists mostly of light-sensing cells, but one tenth of a millimeter of its thickness is populated by image-processing circuitry that is capable of detecting edges (boundaries between light and dark) and motion for about a million tiny image regions. Each of these regions is associated with its own fiber in the optic nerve, and each performs about 10 detections of an edge or a motion each second. The results flow deeper into the brain along the associated fiber.

From long experience working on robot vision systems, I know that similar edge or motion detection, if performed by efficient software, requires the execution of at least 100 computer instructions. Thus, to accomplish the retina's 10 million detections per second would require at least 1,000 MIPS.

The entire human brain is about 75,000 times heavier than the 0.02 gram of processing circuitry in the retina, which implies that it would take, in round numbers, 100 million MIPS (100 trillion instructions per second) to emulate the 1,500-gram human brain. Personal computers in 1999 beat certain insects but lost to the human retina and even to the 0.1-gram brain of a goldfish. A typical PC would have to be at least a million times more powerful to perform like a human brain.

## Brainpower and Utility

Though dispiriting to artificial-intelligence experts, the huge deficit does not mean that the goal of a humanlike artificial brain is unreachable. Computer power for a given price doubled each year in the 1990s, after doubling every 18 months in the 1980s, and every two years before that. Prior to 1990 this progress made possible a great decrease in the cost and size of robot-controlling computers. Cost went from many millions of dollars to a few thousand, and size went from room-filling to handheld. Power, meanwhile, held steady at about 1 MIPS. Since 1990 cost and size reductions have abated, but power has risen to near 1,000 MIPS per computer. At the present pace, only about 30 or 40 years will be needed to close the millionfold gap. Better yet, useful robots don't need full human-scale brainpower.

Commercial and research experiences convince me that the mental power of a guppy—about 1,000 MIPS—will suffice to guide mobile utility robots reliably through unfamiliar surroundings, suiting them for jobs in hundreds of thousands of industrial locations and eventually hundreds of millions of homes. Such machines are less than a decade away but have been elusive for so long that only a few dozen small research groups are now pursuing them.

Commercial mobile robots—the smartest to date, barely insectlike at 10 MIPS—have found few jobs. A paltry 10,000 work worldwide, and the companies that made them are struggling or defunct. (Makes of robot manipulators are not doing much better.) The largest class of commercial mobile robots, known as Automatic Guided Vehicles (AGVs), transport materials in factories and warehouses. Most follow buried signal-emitting wires and detect end points and collisions with switches, a technique developed in the 1960s.

It costs hundreds of thousands of dollars to install guide wires under concrete floors, and the routes are then fixed, making the robots economical only for large, exceptionally stable factories. Some robots made possible by the advent of microprocessors in the 1980s track softer cues, like magnets or optical patterns in tiled floors, and use ultrasonics and infrared proximity sensors to detect and negotiate their way around obstacles.

The most advanced industrial mobile robots, developed since the late 1980s, are guided by occasional navigational markers—for instance, laser-sensed bar codes—and by preexisting features such as walls, corners and doorways. The costly labor of laying guide wires is replaced by custom software that is carefully tuned for each route segment. The small companies that developed the robots discovered many industrial customers eager to automate transport, floor cleaning, security patrol and other routine jobs. Alas, most buyers lost interest as they realized that installation and route changing required time-consuming and expensive work by experienced route programmers of inconsistent availability. Technically successful, the robots fizzled commercially.

In failure, however, they revealed the essentials for success. First, the physical vehicles for various jobs must be reasonably priced. Fortunately, existing AGVs, forklift trucks, floor scrubbers and other industrial machines designed for human riders or for following guide wires can be adapted for autonomy. Second, the customer should not have to call in specialists to put a robot to work or to change its routine; floor cleaning and other mundane tasks cannot bear the cost, time and uncertainty of expert installation. Third, the robots must work reliably for at least six months before encountering a problem or a situation requiring downtime for reprogramming or other alterations. Customers routinely rejected robots that after a month of flawless operation wedged themselves in corners, wandered away lost, rolled over employees' feet or fell down stairs. Six months, though, earned the machines a sick day.

Robots exist that have worked faultlessly for years, perfected by an iterative process that fixes the most frequent failures, revealing successively rarer problems that are corrected in turn. Unfortunately, that kind of reliability has been achieved only for prearranged routes. An insectlike 10 MIPS is just enough to track a few handpicked landmarks on each segment of a robot's path. Such robots are easily confused by minor surprises such as shifted bar codes or blocked corridors (not unlike ants thrown off a scent trail or a moth that has mistaken a streetlight for the Moon).

## A Sense of Space

Robots that chart their own routes emerged from laboratories worldwide in the mid-1990s, as microprocessors reached 100 MIPS. Most build two-dimensional maps from sonar or laser rangefinder scans to locate and route themselves, and the best seem able to navigate office hallways for days before becoming disoriented. Of course, they still fall far short of the six-month commercial criterion. Too often different locations in the coarse maps resemble one another. Conversely, the same location, scanned at different heights, looks different, or small obstacles or awkward protrusions are overlooked. But sensors, computers and techniques are improving, and success is in sight.

My small laboratory is in the race. In the 1980s we devised a way to distill large amounts of noisy sensor data into reliable maps by accumulating statistical evidence of emptiness or occupancy in each cell of a grid representing the surroundings. The approach worked well in two dimensions and guides many of the robots described above.

Three-dimensional maps, 1,000 times richer, promised to be much better but for years seemed computationally out of reach. In 1992 we used economies of scale and other tricks to reduce the costs of three-dimensional maps 100-fold. We now have a test program that accumulates thousands of measurements from stereoscopic camera glimpses to map a room's volume down to centimeter-scale. With 1,000 MIPS, the program digests over a glimpse per second, adequate for slow indoor travel.

## Robot, Version 1.0

One thousand MIPS is only now appearing in high-end desktop PCs. In a few years it will be found in laptops and similar smaller, cheaper computers fit for robots. To prepare for that day, we recently began an intensive three-year project to develop a prototype for commercial products based on such a computer. We plan to automate the learning process to optimize hundreds of evidence-weighing parameters and to write programs to find clear paths, locations, floors, walls, doors and other objects in the three-dimensional maps. We will also test programs that orchestrate the basic capabilities into larger tasks, such as delivery, floor cleaning and security patrol.

The initial testbed will be a small camera-studded mobile robot. Its intelligence will come from two computers: an Apple iBook laptop on board the robot, and an off-board Apple G4-based machine with about 1,000 MIPS that will communicate wirelessly with the iBook. Tiny mass-produced digital camera chips promise to be the cheapest way to get the millions of measurements needed for dense maps.

As a first commercial product, we plan a basketball-size "navigation head" for retrofit onto existing industrial vehicles. It would have multiple stereoscopic cameras, generic software for mapping, recognition and control, a different program for its specific application (such as floor cleaning), and a hardware connection to vehicle power, controls and sensors. Head-equipped vehicles with transport or patrol programs could be taught new routes simply by leading them through once. Floor-cleaning programs would be shown the boundaries of their work area.

Introduced to a job location, the vehicles would understand their changing surroundings competently enough to work at least six months without debilitating mistakes. Ten thousand AGVs, 100,000 cleaning machines and, possibly, a million forklift trucks are candidates for retrofit, and robotization may greatly expand those markets.

## Fast Replay

Income and experience from spatially aware industrial robots would set the stage for smarter yet cheaper ($1,000 rather than $10,000) consumer products, starting probably with small robot vacuum cleaners that automatically learn their way around a home, explore unoccupied rooms and clean whenever needed. I imagine a machine low enough to fit under some furniture, with an even lower extendable brush, that returns to a docking station to recharge and disgorge its dust load. Such machines could open a true mass market for robots.

Commercial success will provoke competition and accelerate investment in manufacturing, engineering investment in manufacturing, engineering and research. Vacuuming robots ought to beget smarter cleaning robots with dusting, scrubbing and picking-up arms, followed by larger multifunction utility robots with stronger, more dexterous arms and better sensors. Programs will be written to make such machines pick up clutter, store, retrieve and deliver things, take inventory, guard homes, open doors, mow lawns, play games and so on. New applications will expand the market and spur further advances when robots fall short in acuity, precision, strength, reach, dexterity, skill or processing power. Capability, numbers sold, engineering and manufacturing quality, and cost-effectiveness will increase in a mutually reinforcing spiral. Perhaps by 2010 the process will have produced the first broadly competent "universal robots," as big as people but with lizardlike 5,000-MIPS minds that can be programmed for almost any simple chore.

Like competent but instinct-ruled reptiles, first-generation universal robots will handle only contingencies explicitly covered in their application programs. Unable to adapt to changing circumstances, they will often perform inefficiently or not at all. Still, so much physical work awaits them in businesses, streets, fields and homes that robotics could begin to overtake pure information technology commercially.

A second generation of universal robot with a mouselike 100,000 MIPS will adapt as the first generation does not and will even be trainable. Besides application programs, such robots would host a suite of software "conditioning modules" that would generate positive and negative reinforcement signals in predefined circumstances. For example, doing jobs fast and keeping its batteries charged will be positive; hitting or breaking something will be negative. There will be other ways to accomplish each stage of an application program, from the minutely specific (grasp the handle underhand or overhand) to the broadly general (work indoors or outdoors). As jobs are repeated, alternatives that result in positive reinforcement will be favored, those with negative outcomes shunned. Slowly but surely, second-generation robots will work increasingly well.

A monkeylike five million MIPS will permit a third generation of robots to learn very quickly from mental rehearsals in simulations that model physical, cultural and psychological factors. Physical properties include shape, weight, strength, texture and appearance of things, and how to handle them. Cultural aspects include a thing's name, value, proper location and purpose. Psychological factors, applied to humans and robots alike, include goals, beliefs, feelings and preferences. Developing the simulators will be a huge undertaking involving thousands of programmers and experience-gathering robots. The simulation would track external events and tune its models to keep them faithful to reality. It would let a robot learn a skill by imitation and afford a kind of consciousness. Asked why there are candles on the table, a third-generation robot might consult its simulation of house, owner and self to reply that it put them there because its owner likes candlelit dinners and it likes to please its owner. Further queries would elicit more details about a simple inner mental life concerned only with concrete situations and people in its work area.

Fourth-generation universal robots with a humanlike 100 million MIPS will be able to abstract and generalize. They will result from melding powerful reasoning programs to third-generation machines. These reasoning programs will be the far more sophisticated descendants of today's theorem provers and expert systems, which mimic human reasoning to make medical diagnoses, schedule routes, make financial decisions, configure computer systems, analyze seismic data to locate oil deposits and so on.

Properly educated, the resulting robots will become quite formidable. In fact, I am sure they will outperform us in any conceivable area of endeavor, intellectual or physical. Inevitably, such a development will lead to a fundamental restructuring of our society. Entire corporations will exist without any human employees or investors at all. Humans will play a pivotal role in formulating the intricate complex of laws that will govern corporate behavior. Ultimately, though, it is likely that our descendants will cease to work in the sense that we do now. They will probably occupy their days with a variety of social, recreational and artistic pursuits, not unlike today's comfortable retirees or the wealthy leisure classes.

The path I've outlined roughly recapitulates the evolution of human intelligence—but 10 million times more rapidly. It suggests that robot intelligence will surpass our own well before 2050. In that case, mass-produced, fully educated robot scientists working diligently, cheaply, rapidly and increasingly effectively will ensure that most of what science knows in 2050 will have been discovered by our artificial progeny!

*Could robots serve as a receptacle for the human mind once the body that housed it has worn down and grown weary with years? Marvin Minsky, casting his imagination into the future, thinks so, and in doing so, he challenges the human family to ponder just what the next step in human evolution might be.*

# Will Robots Inherit the Earth?

## Marvin L. Minsky

Everyone wants wisdom and wealth. Nevertheless, our health often gives out before we achieve them. To lengthen our lives and improve our minds, we will need to change our bodies and brains. To that end, we first must consider how traditional Darwinian evolution brought us to where we are. Then we must imagine ways in which novel replacements for worn body parts might solve our problems of failing health. Next we must invent strategies to augment our brains and gain greater wisdom. Eventually, using nano-technology, we will entirely replace our brains. Once delivered from the limitations of biology, we will decide the length of our lives—with the option of immortality—and choose among other, unimagined capabilities as well.

In such a future, attaining wealth will be easy; the trouble will be in controlling it. Obviously, such changes are difficult to envision, and many thinkers still argue that these advances are impossible, particularly in the domain of artificial intelligence. But the sciences needed to enact this transition are already in the making, and it is time to consider what this new world will be like.

Such a future cannot be realized through biology. In recent times we have learned much about health and how to maintain it. We have devised thousands of specific treatments for specific diseases and disabilities. Yet we do not seem to have increased the maximum length of our life span. Benjamin Franklin lived for 84 years, and except in popular legends and myths no one has ever lived twice that long. According to the estimates of Roy L. Walford, professor of pathology at the University of California at Los Angeles School of Medicine, the average human lifetime was about 22 years in ancient Rome, was about 50 in the developed countries in 1900, and today stands at about 75 in the U.S. Despite this increase, each of those curves seems to terminate sharply near 115 years. Centuries of improvements in health care have had no effect on that maximum.

Why are our life spans so limited? The answer is simple: natural selection favors the genes of those with the most descendants. Those numbers tend to grow exponentially with the number of generations, and so natural selection prefers the genes of those who reproduce at earlier ages. Evolution does not usually preserve genes that lengthen lives beyond that amount adults need to care for their young. Indeed, it may even favor offspring who do not have to compete with living parents. Such competition could promote the accretion of genes that cause death. For example, after spawning, the Mediterranean octopus promptly stops eating and starves itself. If a certain gland is removed, the octopus continues to eat and lives twice as long. Many other animals are programmed to die soon after they cease reproducing. Exceptions to this phenomenon include animals such as ourselves and elephants, whose progeny learn a great deal from the social transmission of accumulated knowledge.

We humans appear to be the longest-lived warm-blooded animals. What selective pressure might have led to our present longevity, which is almost twice that of our other primate relatives? The answer is related to wisdom. Among all mammals our infants are the most poorly equipped to survive by themselves. Perhaps we need not only parents but grandparents, too, to care for us and to pass on precious survival tips.

Even with such advice there are many causes of mortality to which we might succumb. Some deaths result from infections. Our immune systems have evolved versatile ways to cope with most such diseases. Unhappily, those very same immune systems often injure us by treating various parts of ourselves as though they, too, were infectious invaders. This auto-immune blindness leads to diseases such as diabetes, multiple sclerosis, rheumatoid arthritis and many others.

We are also subject to injuries that our bodies cannot repair: accidents, dietary imbalances, chemical poisons, heat, radiation and sundry other influences can deform or chemically alter the molecules of our cells so that they are unable to function. Some of these errors get corrected by replacing defective molecules. Nevertheless, when the replacement rate is too low, errors build up. For example, when the proteins of the eyes' lenses lose their elasticity, we lose our ability to focus and need bifocal spectacles—a Franklin invention.

The major natural causes of death stem from the effects of inherited genes. These genes include those that seem to be largely responsible for heart disease and cancer, the two biggest causes of mortality, as well as countless other disorders, such as cystic fibrosis and sickle cell anemia. New technologies should be able to prevent some of these disorders by replacing those genes.

Most likely, senescence is inevitable in all biological organisms. To be sure, certain species (including some varieties of fish, tortoises and lobsters) do not appear to show any systematic increase in mortality as they age. These animals seem to die mainly from external causes, such as predators or starvation. All the same, we have no records of animals that have lived for as long as 200 years—although this lack does not prove that none exist. Walford and many others believe a carefully designed diet, one seriously restricted in calories, can significantly increase a human's life span but cannot ultimately prevent death.

By learning more about our genes, we should be able to correct or at least postpone many conditions that still plague our later years. Yet even if we found a cure for each specific disease, we would still have to face the general problem of "wearing out" The normal function of every cell involves thousands of chemical processes, each of which sometimes makes random mistakes. Our bodies use many kinds of correction techniques, each triggered by a specific type of mistake. But those random errors happen in so many different ways that no low-level scheme can correct them all.

The problem is that our genetic systems were not designed for very long term maintenance. The relation between genes and cells is exceedingly indirect; there are no blueprints or maps to guide our genes as they build or rebuild the body. To repair defects on larger scales, a body would need some kind of catalogue that specified which types of cells should be located where. In computer programs it is easy to install such redundancy. Many computers maintain unused copies of their most critical system programs and routinely check their integrity. No animals have evolved similar schemes, presumably because such algorithms cannot develop through natural selection. The trouble is that error correction would stop mutation, which would ultimately slow the rate of evolution of an animal's descendants so much that they would be unable to adapt to changes in their environments.

Could we live for several centuries simply by changing some number of genes? After all, we now differ from our relatives, the gorillas and chimpanzees, by only a few thousand genes— and yet we live almost twice as long. If we assume that only a small fraction of those new genes caused that increase in life span, then perhaps no more than 100 or so of those genes were involved. Even if this turned out to be true, though, it would not guarantee that we could gain another century by changing another 100 genes. We might need to change just a few of them—or we might have to change a good many more.

Making new genes and installing them are slowly becoming feasible. But we are already exploiting another approach to combat biological wear and tear: replacing each organ that threatens to fail with a biological or artificial substitute. Some replacements are already routine. Others are on the horizon. Hearts are merely clever pumps. Muscles and bones are motors and beams. Digestive systems are chemical reactors. Eventually, we will find ways to transplant or replace all these parts.

But when it comes to the brain, a transplant will not work. You cannot simply exchange your brain for another and remain the same person. You would lose the knowledge and the processes that constitute your identity. Nevertheless, we might be able to replace certain worn-out parts of brains by transplanting tissue-cultured fetal cells. This procedure would not restore lost knowledge, but that might not matter as much as it seems. We probably store each fragment of knowledge in several different places, in different forms. New parts of the brain could be retrained and reintegrated with the rest, and some of that might even happen spontaneously.

Even before our bodies wear out, I suspect that we often run into limitations in our brains' abilities. As a species, we seem to have reached a plateau in our intellectual development. There is no sign that we are getting smarter. Was Albert Einstein a better scientist than Isaac Newton or Archimedes? Has any playwright in recent years topped William Shakespeare or Euripides? We have learned a lot in 2,000 years, yet much ancient wisdom still seems sound, which makes me think we have not been making much progress. We still do not know how to resolve conflicts between individual goals and global interests. We are so bad at making important decisions that, whenever we can, we leave to chance what we are unsure about.

Why is our wisdom so limited? Is it because we do not have the time to learn very much or that we lack enough capacity? Is it because, according to popular accounts, we use only a fraction of our brains? Could better education help? Of course, but only to a point. Even our best prodigies learn no more than twice as quickly as the rest. Everything takes us too long to learn because our brains are so terribly slow. It would certainly help to have more time, but longevity is not enough. The brain, like other finite things, must reach some limits to what it can learn. We do not know what those limits are; perhaps our brains could keep learning for several more centuries. But at some point, we will need to increase their capacity.

The more we learn about our brains, the more ways we will find to improve them. Each brain has hundreds of specialized regions. We know only a little about what each one does or how it does it, but as soon as we find out how any one part works, researchers will try to devise ways to extend that part's capacity. They will also conceive of entirely new abilities that biology has never provided. As these inventions grow ever more prevalent, we will try to connect them to our brains, perhaps through millions of microscopic electrodes inserted into the great nerve bundle called the corpus callosum, the largest databus in the brain. With further advances, no part of the brain will be out-of-bounds for attaching new accessories. In the end, we will find ways to replace every part of the body and brain and thus repair all the defects and injuries that make our lives so brief.

Needless to say, in doing so we will be making ourselves into machines. Does this mean that machines will replace us? I do not feel that it makes much sense to think in terms of "us" and "them." I much prefer the attitude of Hans P. Moravec of Carnegie Mellon University, who suggests that we think of these future intelligent machines as our own "mind-children."

In the past we have tended to see ourselves as a final product of evolution, but our evolution has not ceased. Indeed, we are now evolving more rapidly, though not in the familiar, slow Darwinian way. It is time that we started to think about our new emerging identities. We can begin to design systems based on inventive kinds of "unnatural selection" that can advance explicit plans and goals and can also exploit the inheritance of acquired characteristics. It took a century for evolutionists to train themselves to avoid such ideas—biologists call them "teleological" and "Lamarckian"—but now we may have to change those rules.

Almost all the knowledge we amass is embodied in various networks inside our brains. These networks consist of huge numbers of tiny nerve cells and smaller structures, called synapses, that control how signals jump from one nerve cell to another. To make a replacement of a human brain, we would need to know something about how each of the synapses relates to the two cells it joins. We would also have to know how each of those structures responds to the various electric fields, hormones, neurotransmitters, nutrients and other chemicals that are active in its neighborhood. A human brain contains trillions of synapses, so this is no small requirement.

Fortunately, we would not need to know every minute detail. If details were important, our brains would not work in the first place. In biological organisms, each system has generally evolved to be insensitive to most of what goes on in the smaller subsystems on which it depends. Therefore, to copy a functional brain it should suffice to replicate just enough of the function of each part to produce its important effects on other parts.

Suppose we wanted to copy a machine, such as a brain, that contained a trillion components. Today we could not do such a thing (even with the necessary knowledge) if we had to build each component separately. But if we had a million construction machines that could each build 1,000 parts per second, our task would take mere minutes. In the decades to come, new fabrication machines will make this possible. Most present-day manufacturing is based on shaping bulk materials. In contrast, nanotechnologists aim to build materials and machinery by placing each atom and molecule precisely where they want it.

By such methods we could make truly identical parts and thus escape from the randomness that hinders conventionally made machines. Today, for example, when we try to etch very small circuits, the sizes of the wires vary so much that we cannot predict their electrical properties. If we can locate each atom exactly, however, the behavior of those wires would be indistinguishable. This capability would lead to new kinds of materials that current techniques could never make; we could endow them with enormous strength or novel quantum properties. These products in turn could lead to computers as small as synapses, having unparalleled speed and efficiency.

Once we can use these techniques to construct a general-purpose assembly machine that operates on atomic scales, further progress should be swift. If it took one week for such a machine to make a copy of itself, we could have a billion copies in less than a year. These devices would transform our world. For example, we could program them to fabricate efficient solar-energy collecting devices and attach these to nearby surfaces. Hence, the devices could power themselves. We would be able to grow fields of microfactories in much the same way that we now grow trees. In such a future we will have little trouble attaining wealth; our trouble will be in learning how to control it. In particular, we must always take care to maintain control over those things (such as ourselves) that might be able to reproduce themselves.

If we want to consider augmenting our brains, we might first ask how much a person knows today. Thomas K. Landauer of Bellcore reviewed many experiments in which people were asked to read text, look at pictures and listen to words, sentences, short passages of music and nonsense syllables. They were later tested to see how much they remembered. In none of these situations were people able to learn, and later remember for any extended period, more than about two bits per second. If one could maintain that rate for 12 hours every day for 100 years, the total would be about three billion bits—less than what we can currently store on a regular five-inch compact disc. In a decade or so that amount should fit on a single computer chip.

Although these experiments do not much resemble what we do in real life, we do not have any hard evidence that people can learn more quickly. Despite common reports about people with "photographic memories," no one seems to have mastered, word for word, the contents of as few as 100 books or of a single major encyclopedia. The complete works of Shakespeare come to about 130 million bits. Landauer's limit implies that a person would need at least four years to memorize them. We have no well-founded estimates of how much information we require to perform skills such as painting or skiing, but I do not see any reason why these activities should not be similarly limited.

The brain is believed to contain on the order of 100 trillion synapses, which should leave plenty of room for those few billion bits of reproducible memories. Someday, using nanotechnology, it should be feasible to build that much storage space into a package as small as a pea.

Once we know what we need to do, our nanotechnologies should enable us to construct replacement bodies and brains that will not be constrained to work at the crawling pace of "real time." The events in our computer chips already happen millions of times faster than those in brain cells. Hence, we could design our "mind-children" to think a million times faster than we do. To such a being, half a minute might seem as long as one of our years and each hour as long as an entire human lifetime.

But could such beings really exist? Many scholars from a

variety of disciplines firmly maintain that machines will never have thoughts like ours, because no matter how we build them, they will always lack some vital ingredient. These thinkers refer to this missing essence by various names: sentience, consciousness, spirit or soul. Philosophers write entire books to prove that because of this deficiency, machines can never feel or understand the kinds of things that people do. Yet every proof in each of those books is flawed by assuming, in one way or another, what it purports to prove—the existence of some magical spark that has no detectable properties. I have no patience with such arguments. We should not be searching for any single missing part. Human thought has many ingredients, and every machine that we have ever built is missing dozens or hundreds of them! Compare what computers do today with what we call "thinking." Clearly, human thinking is far more flexible, resourceful and adaptable. When anything goes even slightly wrong within a present-day computer program, the machine will either come to a halt or generate worthless results. When a person thinks, things are constantly going wrong as well, yet such troubles rarely thwart us. Instead we simply try something else. We look at our problem differently and switch to another strategy. What empowers us to do this?

On my desk lies a textbook about the brain. Its index has approximately 6,000 lines that refer to hundreds of specialized structures. If you happen to injure some of these components, you could lose your ability to remember the names of animals. Another injury might leave you unable to make long-range plans. Another impairment could render you prone to suddenly utter dirty words because of damage to the machinery that normally censors that type of expression. We know from thousands of similar facts that the brain contains diverse machinery. Thus, your knowledge is represented in various forms that are stored in different regions of the brain, to be used by different processes. What are those representations like? We do not yet know.

But in the field of artificial intelligence, researchers have found several useful means to represent knowledge, each better suited to some purposes than to others. The most popular ones use collections of "if-then" rules. Other systems use structures called frames, which resemble forms that are to be filled out. Yet other programs use web-like networks or schemes that resemble trees or sequences of planlike scripts. Some systems store knowledge in languagelike sentences or in expressions of mathematical logic. A programmer starts any new job by trying to decide which representation will best accomplish the task at hand. Typically a computer program uses a single representation, which, should it fail, can cause the system to break down. This shortcoming justifies the common complaint that computers do not really "understand" what they are doing.

What does it mean to understand? Many philosophers have declared that understanding (or meaning or consciousness) must be a basic, elemental ability that only a living mind can possess. To me, this claim appears to be a symptom of "physics envy"—that is, they are jealous of how well physical science has explained so much in terms of so few principles. Physicists have done very well by rejecting all explanations that seem too complicated and then searching instead for simple ones. Still, this method does not work when we are addressing the full complexity of the brain. Here is an abridgment of what I said about the ability to understand in my book The Society of Mind: If you understand something in only one way, then you do not really understand it at all. This is because if something goes wrong you get stuck with a thought that just sits in your mind with nowhere to go. The secret of what anything means to us depends on how we have connected it to all the other things we know. This is why, when someone learns "by rote," we say that they do not really understand. However, if you have several different representations, when one approach fails you can try another. Of course, making too many indiscriminate connections will turn a mind to mush. But well-connected representations let you turn ideas around in your mind, to envision things from many perspectives, until you find one that works for you. And that is what we mean by thinking!

I think flexibility explains why, at the moment, thinking is easy for us and hard for computers. In The Society of Mind, I suggest that the brain rarely uses a single representation. Instead it always runs several scenarios in parallel so that multiple viewpoints are always available. Furthermore, each system is supervised by other, higher-level ones that keep track of their performance and reformulate problems when necessary. Because each part and process in the brain may have deficiencies, we should expect to find other parts that try to detect and correct such bugs.

In order to think effectively, you need multiple processes to help you describe, predict, explain, abstract and plan what your mind should do next. The reason we can think so well is not because we house mysterious sparklike talents and gifts but because we employ societies of agencies that work in concert to keep us from getting stuck. When we discover how these societies work, we can put them inside computers, too. Then if one procedure in a program gets stuck, another might suggest an alternative approach. If you saw a machine do things like that, you would certainly think it was conscious.

This article bears on our rights to have children, to change our genes and to die if we so wish. No popular ethical system yet, be it humanist or religion-based, has shown itself able to face the challenges that already confront us. How many people should occupy the earth? What sorts of people should they be? How should we share the available space? Clearly, we must change our ideas about making additional children. Individuals now are conceived by chance. Someday, instead, they could be "composed" in accord with considered desires and designs. Furthermore, when we build new brains, these need not start out the way ours do, with so little knowledge about the world.

What kinds of things should our "mind-children" know? How many of them should we produce, and who should decide their attributes?

Traditional systems of ethical thought are focused mainly on individuals, as though they were the only entities of value. Obviously, we must also consider the rights and the roles of larger-scale beings—such as the superpersons we term cultures and the great, growing systems called sciences—that help us understand the world. How many such entities do we want? Which are the kinds that we most need? We ought to be wary of ones that get locked into forms that resist all further growth. Some future options have never been seen: imagine a scheme that could review both your mentality and mine and then compile a new, merged mind based on that shared experience.

Whatever the unknown future may bring, we are already changing the rules that made us. Most of us will fear change, but others will surely want to escape from our present limitations. When I decided to write this article, I tried these ideas out on several groups. I was amazed to find that at least three quarters of the individuals with whom I spoke seemed to feel our life spans were already too long. "Why would anyone want to live for 500 years? Wouldn't it be boring? What if you outlived all your friends? What would you do with all that time?" they asked. It seemed as though they secretly feared that they did not deserve to live so long. I find it rather worrisome that so many people are resigned to die. Might not such people, who feel that they do not have much to lose, be dangerous?

My scientist friends showed few such concerns. "There are countless things that I want to find out and so many problems I want to solve that I could use many centuries," they said. Certainly immortality would seem unattractive if it meant endless infirmity, debility and dependency on others, but we are assuming a state of perfect health. Some people expressed a sounder concern—that the old ones must die because young ones are needed to weed out their worn-out ideas. Yet if it is true, as I fear, that we are approaching our intellectual limits, then that response is not a good answer. We would still be cut off from the larger ideas in those oceans of wisdom beyond our grasp.

Will robots inherit the earth? Yes, but they will be our children. We owe our minds to the deaths and lives of all the creatures that were ever engaged in the struggle called evolution. Our job is to see that all this work shall not end up in meaningless waste.

*Questioning everything, leaving no assumption unharried, Marvin Minsky would no doubt find a soulmate in the ancient Greek philosopher Socrates. Minsky's restless intellect gave birth to the modern branch of artificial intelligence, and yet, he stands as one of the most vocal and potent critics of that science. John Horgan spoke with Minsky and presents a profile of a multidimensional personality whose interests range from the lofty peaks of philosophical inquiry to the design of a new ride for Disney World.*

# Marvin L. Minsky: The Mastermind of Artificial Intelligence

## John Horgan

**M**arvin Minsky's ideas about the mind may—or may not—offer lasting insights. But they certainly reveal much about the mind of Minsky. According to Minsky, the mind is not a unified entity but a "society" of elements that both complement and compete with one another. Minsky's emphasis on multiplicity seems to transcend science; he views single-mindedness with a kind of horror. "If there's something you like very much, then you should regard this not as you feeling good but as a kind of brain cancer," he explains, "because it means that some small part of your mind has figured out how to turn off all the other things."

Minsky even recoils at the tendency of ordinary mortals, once they have invested the time in learning to do something, to keep doing it. To counter this trait, which he calls the investment principle, Minsky has trained himself to "enjoy the feeling of awkwardness" aroused by confronting an entirely new problem. "It's so thrilling not to be able to do something," he remarks.

This credo has served Minsky well in his role as a founding

father of artificial intelligence. Called AI by insiders, the field is dedicated to the proposition that brains are nothing more than machines, albeit extremely complicated ones, whose abilities will someday be duplicated by computers. In pursuit of the goals of AI, Minsky has drawn on computer science, robotics, mathematics, neuroscience, psychology, philosophy and even science fiction. His ideas have in turn influenced all these fields as well as AI itself.

But the same traits that made Minsky a successful pioneer of AI have led him to become increasingly alienated from the field as it matures. Before my meeting with Minsky, in fact, other AI workers warn me that he might be somewhat cranky; if I do not want the interview cut short, I should not ask him too directly about the current slump in AI or what some workers characterize as his own waning influence in the field. One prominent theorist pleads with me not to take advantage of Minsky's penchant for hyperbole. "Ask him if he means it, and if he doesn't say it three times, you shouldn't use it," the theorist urges.

Minsky is rather edgy when I meet him in his office at the Artificial Intelligence Laboratory. He fidgets ceaselessly, blinking, waggling his foot, pushing things about his desk. Unlike most scientific celebrities, he gives the impression of conceiving ideas and tropes from scratch rather than retrieving them from memory. He is often but not always incisive. "I'm rambling here," he says glumly after a riff on the nature of verification in AI collapses in a heap of sentence fragments.

Even his physical appearance has an improvisational air. His large, round head seems entirely bald but is actually fringed by hairs as transparent as optical fibers. He wears a crocheted belt that supports, in addition to his pants, a belly pack and a holster containing pliers with retractable jaws. With his paunch and vaguely Asian features, he resembles a high-tech Buddha.

Minsky is unable, or unwilling, to inhabit any emotion for long. Early on, as predicted, he plays the curmudgeon. His only rival in grasping the mind's complexity is dead: "Freud has the best theories so far, next to mine, of what it takes to make a mind," Minsky declares. If AI has not progressed as fast as it should have, that is because modern researchers have succumbed to "physics envy"—the desire to reduce the intricacies of the brain to simple formulae—and to the dreaded investment principle. "They are defining smaller and smaller subspecialities that they examine in more detail, but they're not open to doing things in a different way." Even M.I.T.'s own AI lab, which he founded, is guilty. "I don't consider this to be a serious research institution at the moment," he sneers.

But a metamorphosis occurs when, touring the AI lab, we stop to chat with some researchers in a lounge. Minsky engages in amiable shoptalk about chess-playing computers. He then recounts how the late science fictionist Isaac Asimov always refused Minsky's invitations to see the robots being built at M.I.T. out of fear that his imagination "would be weighed down by this boring realism."

One lounger, noticing that he and Minsky have the same pliers, yanks his instrument from its holster and with a flick of his wrist snaps the retractable jaws into place. "En garde," he says. Minsky, grinning, draws his weapon, and he and his challenger whip their pliers repeatedly at each other, like punks practicing their switchblade technique. Minsky expounds on both the versatility and—an important point for him—the drawbacks of the pliers; his pair pinches him during certain maneuvers. "Can you take it apart with itself?" someone asks. Minsky and his colleagues share a laugh at this reference to a fundamental problem in robotics.

Returning to Minsky's office, we encounter a young, extremely pregnant graduate student. She is scheduled for an oral doctoral exam the next day. "Are you nervous?" Minsky inquires. "A little," she replies. "You shouldn't be," he says and gently touches his forehead to hers. I realize, watching this scene, that there are many Minskys.

Too many, according to Minsky. As a child, the son of a New York City surgeon, he was a prodigy in both mathematics and music. Minsky still occasionally finds himself composing "Bach-like things"—an electric organ crowds his office—but he tries to resist the impulse by convincing himself that music suppresses thought. "I had to kill the musician at some point," he says. "It comes back every now and then, and I hit it."

Minsky started to think about thinking—or, more specifically, about learning—in high school. Although he received undergraduate and graduate degrees in mathematics (from Harvard and Princeton universities, respectively), he scavenged in other disciplines for ideas he felt could illuminate the mind. In 1951 he and a colleague built a machine, made of vacuum tubes, motors and servomechanisms, that could "learn" how to navigate a maze. It was the first neural network ever built. Minsky followed this engineering project with a doctoral thesis on automated learning.

In 1959 he and John McCarthy—who is credited with having coined the term "artificial intelligence"—founded what became the M.I.T. Artificial Intelligence Laboratory. McCarthy left four years later to found his own laboratory at Stanford University, and since then, he and Minsky have had an intellectual parting of the ways. McCarthy has championed AI models based on logic, whereas Minsky contends that logic requires precise definitions that the real world fails to respect. The definition of a bird as a feathered animal that flies, he points out, does not apply if the bird is dead or caged or has had its feet encased in concrete "or has been meditating and decided flying is egotistical."

He has been even harder on neural networks, the technology he helped to nurture. In 1969 he and Seymour Papert of M.I.T. presented a detailed critique of a then popular neural network in a book entitled *Perceptrons*. The book is often said to have dealt neural networks a nearly mortal blow; funding fell off rapidly, and the field languished for more than a decade before it slowly began regenerating. Minsky's intention was not to destroy the field, as some observers have claimed—"That's crazy," he snaps—but to outline the limits of the technology.

Although Minsky applauds the recent resurrection of neural networks, he charges that some "semicommercial" researchers are not as forthcoming as they should be. "They write a paper saying, 'Look, it did this,' and they don't consider it equally wonderful to say, 'Look, it can't do that.' Most of them are not doing good science, because they're hiding the deficiencies." Minsky insists that no single approach can reproduce the intricacies of the mind, because the mind itself employs many fallible methods that back up and check one another. The mind, he muses, is a tractor-trailer, rolling on many wheels, but AI workers "keep designing unicycles."

Some aspects of the mind have proved harder to understand or reproduce than Minsky expected. He confirms the often-told anecdote that in the early 1960s he assigned artificial vision, now recognized as a profoundly difficult problem, to a student as a summer project. But he expects all the major questions in AI to be solved as imaging and electrode techniques reveal the brain in ever finer detail. "Everything we've done up to now I regard as like chemistry before Lavoisier," he remarks.

Minsky poured his thoughts about thinking into *The Society of Mind*, published in 1985. The book consists of 270 essays, most of them only one page long, which range from rather technical discussions of neural wiring to philosophical excursions into the nature of human identity. In the book's prologue, Minsky contended that the work's atomized structure reflects its major theme, that "you can build a mind from many little parts, each mindless by itself." "As far as I know, nobody read the book," Minsky grumbles.

Minsky has nothing but contempt for those who believe that computers, while they may be able to mimic certain aspects of human intelligence, can never become truly conscious.

"They're idiots," he fumes. (Minsky is kinder to me when I make the mistake of suggesting that there might always be a qualitative difference between humans and artificial machines; he calls me a "racist.")

The mystery of consciousness is "trivial," Minsky declares. "I've solved it, and I don't understand why people don't listen." Consciousness, Minsky explains, involves one part of the mind monitoring the behavior of other parts. This function requires little more than short-term memory, or a "low-grade system for keeping records." In fact, computer programs such as LISP, which have memory features that allow their processing steps to be retraced, are "extremely conscious," Minsky asserts— more so than humans, who have pitifully shallow short-term memories.

Like many AI practitioners, Minsky predicts that computers will someday evolve far beyond humans, who are nothing but "dressed-up chimpanzees." Humans may be able to "download" their personalities into computers and thereby become smarter and more reliable. This trick may yield infinite life, among other perks. Minsky envisions making copies of himself that could undergo experiences he would otherwise shun. "I regard religious experience as very risky, because it can destroy your brain. But if I had a backup copy—"

Meanwhile the *Ur*-Minsky remains restless. Hollywood may provide one outlet for his energies. That becomes clear when Laurel, an administrator at the AI lab, sticks her head in the office to ask what's new. Minsky replies that the Disney corporation has hired him to design a "magic carpet ride," based on its hit movie *Aladdin*, for one of its theme parks. Minsky has been working on a virtual-reality scheme at a laboratory Disney has set up for special effects. "I love it," Minsky says of the laboratory. "It's just like the AI lab used to be."

Noting that Stephen W. Hawking, the English cosmologist, appeared on *Star Trek, The Next Generation*, Laurel suggests that Minsky is well suited for playing "an alien genius" on the television show. Evil or benign? I ask. "Oh, either," Laurel replies. Minsky seems intrigued, but he worries that he may be unable to rehearse scenes properly. "I can't say the same thing twice," he confesses.

Minsky is also working on a new book, *The Emotion Machine*. "That's a person," Minsky says of the title. One goal of the book, he notes, is to help people think constructively about thinking. "I'm interested in people who are trying to do some work but keep watching television or going to baseball games." The book will advise such people to make "a little block diagram" of their minds, with different, competing agents labeled. I try to imagine Minsky as a self-help guru, propounding his AI-oriented program on "Donahue." Then, recalling the way he comforted the pregnant graduate student, I think, Why not?

C. Clarke's and film di
y Kubrick's vision of co
logy at the turn of the
s computers are vastly
r, more portable and use
aces that forgo the type
controls found on the s
ery 1, B y and large-

# Conclusion

ould a thinking machine eventually outperform a human being? In some activities deemed intelligent, such as playing chess or computing math problems, they already do. But the benchmark of artificial intelligence-a machine that thinks better, learns better, and perceives better than its human creator-is still a long way off.

If nothing else, the pursuit of artificial intelligence has given us a whole new appreciation for human intelligence. The mere act of attempting to duplicate it has revealed far more about how we think and perceive the world than previously suspected. Thinking, learning, and ordering thoughts in three-dimensional reality require an exquisite array of systems and assumptions about the world. Humans are born with these tools. Machines must acquire them.

As AI advances, society will be forced to address moral and ethical issues associated with artificial intelligence and (at this point, hypothetical) artificial humans, such as, should they be granted "human" rights? A similar situation is already occur-ring. Wildlife scientists and naturalists who wish to establish universal rights for the great apes, our closest living relatives, have initiated the debate. They are pushing for a proclamation of great-ape rights by the United Nations. Will a future proclamation be necessary to free artificial intelligences from human bondage? Should we allow this to happen? Should we fear our electronic offspring? Will ambulatory AI machines proceed down the Terminator's path ... or down the benign road to become helpful non-human assistants? Will artificial humans inherit the planet, as some scientists are now inclined to say, or will the melding of biology and bionics simply necessitate a new definition of "human"? We may not have all the answers yet, but the questions will become more important as each new invention leads toward true artificial awareness.

But perhaps misgivings about artificial intelligence are unfounded. After all, the human geniuses that gave birth to this brave new science were just doing what comes naturally to human beings—asking the right questions and tinkering with machines. AI may be the natural extension of humans as tool-makers, and humanity's future may be inextricably linked with our creations.

# Index

LaVergne, TN USA
07 July 2010
188615LV00001B/21/A